PHOTOFACT®
TELEVISION COURSE

by
The Howard W. Sams Editorial Staff

Howard W. Sams & Co., Inc.
4300 WEST 62ND ST. INDIANAPOLIS, INDIANA 46268 USA

Preface

The television industry has experienced phenomenal growth since the first edition of *PHOTOFACT® Television Course* was introduced thirty years ago. Although the basic concepts and standards remain the same, many of the circuits and components used in present-day television receivers bear little resemblance to their earlier counterparts. This Fifth Edition has been updated to reflect the latest knowledge, skills, and techniques of the industry.

The previous editions of this book have been recognized as one of the most authoritative and comprehensive texts of its type available. It has provided the basic knowledge for thousands who have become skilled in the field of television engineering and servicing. The newest edition continues to provide the fundamental concepts of television receiver design in terms of the technology of today. Much of the book has been revised to assure you of the most accurate, complete, and timely information concerning the basic principles of practical theory and operation of black-and-white television receiver circuits.

If you are just entering the television servicing field, this book should prove invaluable. Even if you are an experienced television technician, you will find this book a useful and authoritative reference source.

HOWARD W. SAMS EDITORIAL STAFF

Contents

INTRODUCTION 7

Section 1 — Beam Formation and Control

CHAPTER 1

CATHODE-RAY TUBE—BEAM FORMATION AND ELECTROSTATIC
CONTROL 11
Beam Formation — Beam Control — Beam Effect on the Fluorescent Screen

CHAPTER 2

CATHODE-RAY TUBE—ELECTROMAGNETIC CONTROL OF THE BEAM . . 19
Effect of Magnetic Fields on an Electron Beam — Beam Focusing — Beam
Deflection — Beam Centering — Removal of Ions from Electron Beam

CHAPTER 3

THE CAMERA TUBE 25
Image Orthicon — Vidicon — Monoscope — Scanning

CHAPTER 4

POWER SUPPLIES 31
Solid-State Rectifiers — Regulation and Filtering

Section 2 — Beam Deflection

CHAPTER 5

RESISTANCE-CAPACITANCE CIRCUIT CHARACTERISTICS 39
RC Circuit Charging — RC Circuit Discharge — Time Constants of an RC
Circuit — Formation of Square and Sawtooth Waves

CHAPTER 6

SAWTOOTH GENERATORS 45
Neon-Tube Oscillators — Thyratron Oscillator — Vacuum-Tube Sawtooth
Generators — Transistor Sawtooth Generators — A Summary of Multivibra-
tors and Blocking Oscillators

CHAPTER 7

SAWTOOTH GENERATOR CONTROL AND PRODUCTION OF
SCANNING WAVEFORMS 59
Control of Scanning Generators by Sync Pulses — Scanning Requirements
for Picture Tubes — Peaking Circuits for Electromagnetic Deflection

CHAPTER 8

DEFLECTION SYSTEMS 67
Horizontal-Deflection Circuits — High-Voltage Power Supply — Horizontal-
Oscillator Circuits— Vertical-Deflection Systems

Section 3 — Beam Modulation and Synchronization

CHAPTER 9

THE COMPOSITE TELEVISION SIGNAL 85
Vestigial-Sideband Video Modulation — The Video Signal — The Direct-
Current Component of the Video Signal

CHAPTER 10

SYNC-PULSE SEPARATION, AMPLIFICATION, AND USE 93
Sync-Pulse Separation — Sync-Pulse Amplification, Clipping, and Shaping
— Sorting of the Individual Horizontal and Vertical Pulses — The Function
of Vertical-Equalizing Pulses — Action of the Horizontal-Differentiating
Circuit During the Vertical Pulse

CHAPTER 11

THE RECEIVING ANTENNA 105
Polarization of the Transmitted Wave — Types of Wave Paths Between the
Transmitter and Receiver — Wavelengths of the Television Channels — The
Noise Problems — Ghosts Due to Multiple-Path Transmission — Ghosts Due
to Reflections in the Lead-In — The Half-Wave Dipole — The Folded Dipole
— Antenna Structures Employing the Dipole With Reflectors and/or Direc-
tors–Yagi Arrays — The Broadband Problem — Stacked Arrays — The Cor-
ner-Reflector Antenna — Antenna Rotators — Types of Lead-In — Television
Reception in Fringe Areas — Master Antenna Systems

CHAPTER 12

TUNERS 123
Tuning Systems — RF Amplifiers — Oscillator Circuits — Mixer Circuits
— Automatic Fine Tuning — UHF Tuners

CHAPTER 13

VIDEO IF AMPLIFIERS AND DETECTORS 139
Introduction — Video IF Systems — Video Detectors

CHAPTER 14

SOUND IF AMPLIFIERS AND AUDIO DETECTORS 151
Sound IF Takeoff — Typical Sound IF Systems — Audio Detectors

CHAPTER 15

VIDEO AMPLIFIERS 159
Bandwidth and Gain Characteristics — Low-Frequency Compensation —
High-Frequency Compensation — Phase Shift in the Video Amplifier —
Smearing of the Picture Due to Amplitude or Phase Distortion — Bright-
ness Control — Contrast Control — Video Coupling to CRT

CHAPTER 16

AUTOMATIC GAIN CONTROL 171
Rectified AGC — Amplified AGC — Keyed AGC — Delayed AGC for RF
Stage

CHAPTER 17

RECEIVER CONTROLS—APPLICATION AND ADJUSTMENT 179
Front-Panel or Viewer-Operated Controls — Preset Controls — Classification
According to Function

APPENDIX

GLOSSARY 189

ANSWERS TO QUESTIONS 197

INDEX 201

Introduction

The principles of television can be understood best by careful analysis of the various sections and stages that comprise the receiver. The basic sections include a tuner for selecting the desired television channel, sound and video circuits, beam formation and control circuits, horizontal and vertical beam-deflection and synchronization circuits, and the power-supply circuits.

The heart of the television receiver is a cathode-ray tube, or picture tube. The cathode-ray tube is essentially a vacuum tube that produces a narrow beam of electrons. These electrons are accelerated by a large positive potential on the anode of the picture tube, and they move rapidly toward the phosphor-coated screen which glows when bombarded by the high-speed electrons. The beam can be deflected and modulated (changed in intensity) so as to produce an image on the face of the picture tube.

Television also requires the use of another distinctively different type of cathode-ray tube—the camera tube. The camera tube converts light energy into electrical energy which represents the video signal. This signal from the camera tube is amplified and used to modulate a carrier wave. After transmission and reception, the picture tube then "reconverts the video signal back into light energy for viewing. Transmitted simultaneously with the video signal are the blanking and synchronizing pulses. These pulses maintain the horizontal and vertical scanning circuits in synchronization with the transmitter to produce 30 frames per second. Each frame consists of 525 horizontal lines according to present-day U.S. standards.

To ease comprehension in the study of television reception, the receiver is divided into three major circuit sections which have independent functions. These sections are:

1. *Beam formation and control*—involves the study of the picture tube, the camera tube, and electron optics. A discussion of the low-voltage power supplies used in television receivers is also included in this section.
2. *Beam-deflection systems*—involves the study of blocking oscillators, free-running multivibrators, and waveshaping circuits. Includes the vertical and horizontal deflection circuits and the development of the high potentials applied to the anode of the picture tube.
3. *Beam modulation and synchronization*—involves the study of high-frequency broadband receivers, video detectors, and video amplifiers.

7

Also includes a discussion of television receiving antennas and receiver controls.

These three master subjects serve as guides as to what and how to study. They are in correct sequence to provide a clear understanding of what takes place in a television receiver. The reader is advised to thoroughly grasp one subject before proceeding to the next.

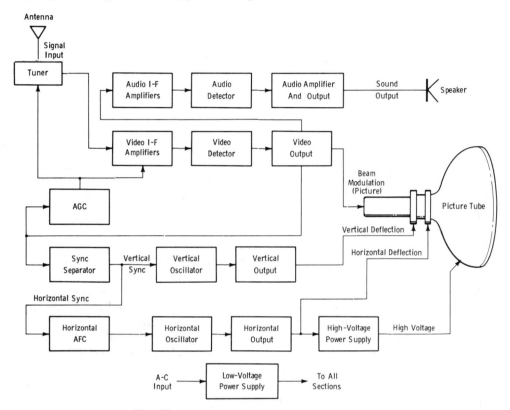

Simplified block diagram of a television receiver.

Beam Formation and Control

CHAPTER 1

Cathode-Ray Tube—Beam Formation and Electrostatic Control

The cathode-ray tube is employed in a number of applications besides television. One such application is the oscilloscope. The cathode-ray tubes used in oscillography and television are similar, and the basic principles apply to both. The oscilloscope tube, however, is simpler and for this reason is used as an introduction to the study of cathode-ray tubes.

BEAM FORMATION

Fig. 1-1 shows a simple cathode-ray tube used in early experiments to develop a narrow beam of electrons. Emitted from a hot filament, the electrons rush forward, attracted by a positively charged plate. Accelerated by this positive charge, the beam reaches a velocity of many thousands of miles per second, the exact speed depending on the force of the positive attraction. The equal negative charges carried by the electrons set up a repelling action which causes the emitted beam to widen or scatter. However, the small hole in the center of the anode narrows the beam and permits some of the fast-moving electrons to race toward the face of the tube without losing much speed.

The electrons comprising this narrow beam are traveling too fast for any effective scattering to take place. Thus, a small illuminated spot is produced on the fluorescent coating on the face (screen) of the tube when it is struck by the beam. The scattered electrons collected by the positively charged anode produce a current through the B supply circuit.

The color of the spot depends on the type of chemical employed on the face of the tube. Oscilloscope tubes generally have a P1 phosphor, which produces a green trace. Television tubes for black-and-white receivers always have a P4 phosphor, which produces a white trace.

Cathode-ray tubes contain the following: (1) an electron-beam source, (2) a fluorescent screen for visible indication, (3) a means for varying the intensity (which controls the brilliance of the spot), (4) a method of focusing the beam, which controls the size of the spot, and (5) provision for deflecting the beam, which controls the position of the spot.

In modern cathode-ray tubes, the electron source is an indirectly heated cathode, or electron emitter. This cathode is a small cylinder of nickel about one-eighth inch in diameter and about one-half inch in length. The nickel sleeve is coated on one end with oxide, which permits a large number of electrons to be emitted toward the fluorescent screen. The heater is a tungsten wire filament wound in the form of a noninductive spiral. The spiral winding tends to cancel any magnetic field that might affect the electron beam. The filament coil is insulated and is inserted into the cathode sleeve. For better heat conduction to the cathode, the insulated filament contacts the nickel sleeve. This important cathode-ray tube element is illustrated in Fig. 1-2.

BEAM CONTROL

Grid-Control Element

The elementary cathode-ray tube illustrated in Fig. 1-1 has no controlling element to limit the number of electrons emitted from the cathode.

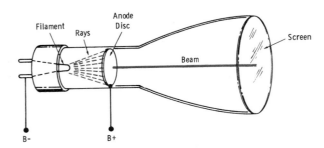

Fig. 1-1. Formation of a narrow beam of electrons in an early experimental cathode-ray tube.

The brilliance of the spot on the screen must be controlled. This requires an additional element, known as the control grid, between the cathode and the positively charged anode. In modern tubes, this element is a metal cylinder completely enclosing the cathode element (Fig. 1-3A). The strategic position of the control grid allows the quantity of electrons in the beam to be controlled. The more negative the control grid is biased with respect to the cathode, the fewer electrons in the beam, and the less the intensity of light produced on the screen. As the control grid is made less negative with respect to the cathode, the spot brilliance is increased. The direction of the electron emission is governed by an aperture in the disc at the end of the grid cylinder.

The lines of force of the electrostatic field, developed by the difference of potential between the cathode and control grid, and their effect on the beam are illustrated in Fig. 1-3B.

If the voltage on the control grid is made more negative with respect to the cathode, fewer electrons will be admitted to the beam. If the control grid is sufficiently negative, the beam will be completely shut off, or blanked out. Also, the negative field associated with the control grid causes the beam to cross over itself after it passes through the control-grid aperture (Fig. 1-3B). This crossover is similar to that of an optical lens and concentrates the beam of electrons into a fine point.

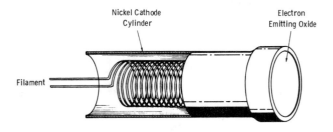

Fig. 1-2. Heater and cathode assembly.

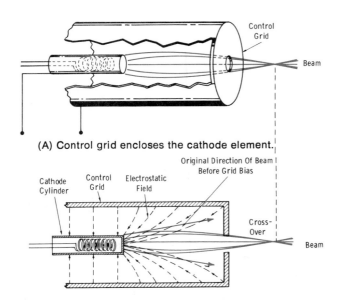

(A) Control grid encloses the cathode element.

(B) Electrostatic lines of force between the cathode and the grid control the beam.

Fig. 1-3. Action of cathode and control-grid assembly.

Hence, the phrase "electron optics" came into use. The concentration of the beam is shown in Fig. 1-3B.

To summarize, we recall that the control grid has three functions:

1. To control the brilliance from zero to maximum at the crt screen.
2. To concentrate the lens action of the beam by effecting a crossover.
3. To provide a means for inserting a varying signal for intensity modulation of the beam.

In oscilloscopes, a potentiometer known as the intensity control varies the control-grid bias and enables the brilliance to be manually adjusted to a comfortable level.

Focus- and Accelerating-Control Elements

So far, we have been able to control the brilliance of the spot, but another control is necessary to bring the spot into sharp focus. The focusing of an electron beam in a cathode-ray tube is similar to the focusing of a light beam (Fig. 1-4). From the previous discussion we learned that the control grid did focus the beam to a point slightly beyond its aperture, but the beam begins to widen again after the crossover point. Therefore, additional focusing is needed. In earlier picture tubes,

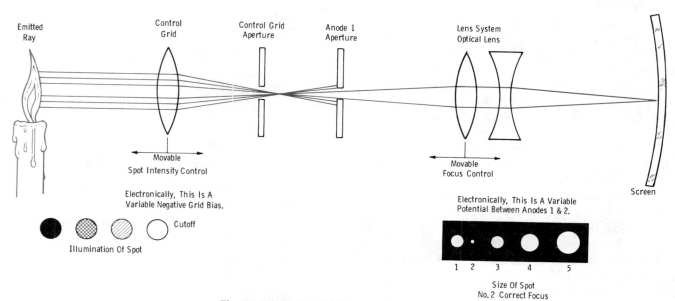

Fig. 1-4. Similarity of light ray and cathode ray.

focusing was accomplished by two cylindrical anodes (Fig. 1-5).

Here, the two anodes were operated at different potentials, and the lines of force between them established the lens action. The moving electrons were subjected to forces as they entered the electrostatic field set up by anodes 1 and 2. To understand clearly what takes place, refer to Fig. 1-5. Here, the electron beam is seen entering the electrostatic field developed by a difference of potential between the two anodes. Coming from point A, the original crossover, the beam enters a new field. As the electrons cross the first static lines (1 and 2 in Fig. 1-5), they tend to change their

course, since the forces acting on them will repel negative charges because of the potential gradient between anodes 1 and 2. Although both anodes are positive with respect to the cathode, anode 1, operating at a much lower potential, is negative with respect to anode 2, and an electrostatic field which tends to bend the path of the electrons entering that field is created.

The electrons, traveling at a high speed, are gradually bent into a beam, the greatest repelling force occurring at a point somewhere between the two anodes. When the electrons of the beam converge near the axis of the tube, the lines of force are running almost parallel to the axis, and the

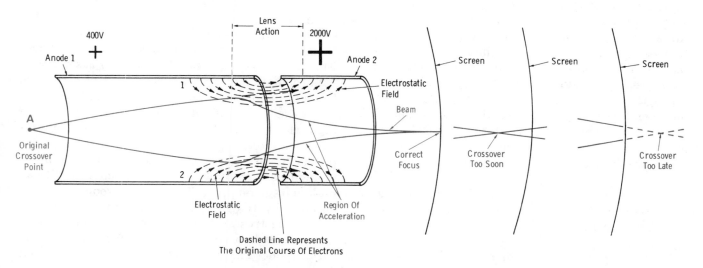

Fig. 1-5. Effect of an electrostatic field on an electron beam.

electrons in the beam begin to accelerate. At the exit end of anode 2, the field is relatively weak, and the electrons keep their direct course, aided by the velocity gained while traveling through anode 2. The two anode cylinders are similar to the grid cylinder, except anode 1 usually is longer and has two aperture discs spaced for better focus.

After the beam has passed through anode 2, it meets at a second crossover point. This second crossover is adjusted to take place as the beam arrives at the surface of the screen.

The focus of an electrostatic cathode-ray tube (Fig. 1-5) is generally adjusted by varying the voltage to anode 1. This controls the amount of force the electrostatic field exerts on the electron beam. By rotating the focus control and observing the screen, the beam can easily be brought into sharp focus.

Before discussing variations in anode structure of the type illustrated in Fig. 1-5, it will be helpful to clarify the terminology regarding these elements.

Since the first cathode-ray tubes had two cylindrical anodes (Fig. 1-5), the terms "first anode" and "second anode" were quite logical. However, the newer tubes have more than two anodes or have physically split anodes, and it is difficult to name them according to their positions in the tube. Therefore, they should be labeled according to their purpose. Consequently, anode 1, which controls the focus, can be called the focus anode, and anode 2, the higher potential element, can be called the accelerating anode.

Fig. 1-6 shows a more recent form of anode assembly in a typical electrostatic tube. The cylindrical anode adjacent to the control grid is an accelerating anode, instead of the focus anode employed in Fig. 1-5. The circular disc with the large aperture, following the accelerating anode, is actually the focus anode. The shorter cylindrical anode, combined with the second circular disc, is electrically connected to the first cylindrical anode and is considered part of the accelerating anode structure. The reasons for this construction are:

1. Removal of the focus anode from its position near the control grid lessens any interaction between intensity and focus control adjustments.
2. By proper placement of the focus anode, its aperture can be made larger. The amount of

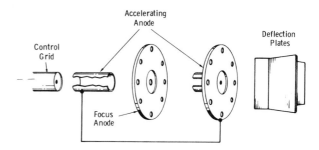

Fig. 1-6. Typical anode assembly.

beam current drawn by this anode will be reduced, and its effect on the beam intensity will be lessened. This is possible because of the separation of portions of the accelerating anode, enabling insertion of the focus anode.

3. In any cathode-ray tube with appreciable focus-anode current, the focus-control circuits must use sufficient bleeder current to ensure reliable operation. With the construction outlined in Fig. 1-6, the requirement for bleeder current can be reduced considerably or eliminated altogether.

Remember that construction of these tubes may vary according to the manufacturers' preferences of electrical design and physical support of the various elements.

Summing up, we know that the focus and accelerating anodes have two functions:

1. Focusing the beam for sharpest detail of the image on the screen. (The degree of focusing is manually controlled.)
2. Accelerating the beam.

The assembly discussed so far constitutes the electron gun, so called because it "shoots" negative particles (electrons) to a screen or target. Fig. 1-7 shows a typical electron gun assembly.

Cathode-ray tubes with combined electrostatic focusing and magnetic deflection are currently being used. The earliest of these tubes were called high electrostatic-focus tubes, because they required approximately 3000 to 5000 dc volts for proper focus. These tubes had to have a high-voltage power supply separate from the regular high-voltage supply. They were soon rendered obsolete by the low electrostatic-focus tubes.

Low electrostatic-focus tubes could be focused by voltages between −100 and +500 volts on the focus anode. These low voltages could be obtained from the low-voltage power supply in the receiver;

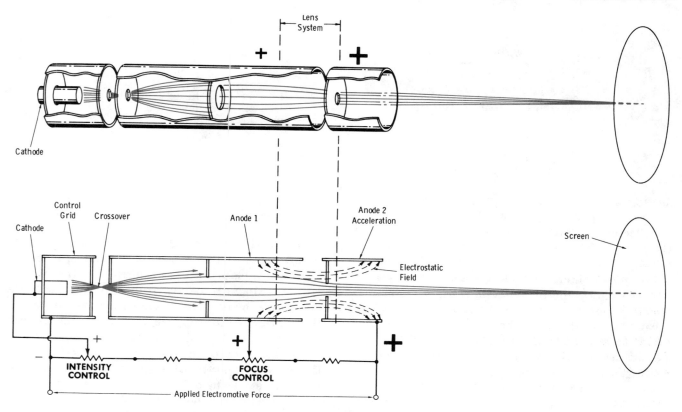

Fig. 1-7. Electron gun assembly.

thus, the extra high-voltage supply could be eliminated. Both the low and the high electrostatic-focus tubes required a potentiometer-type control to vary the focus anode voltage for best focusing. Fig. 1-8 shows the electron gun of a low electrostatic-focus tube.

Several picture tubes having automatic electrostatic focusing were produced. In these tubes, an internal resistance connected the focusing electrode to the cathode of the electron gun. Since the focusing electrode is in or near the electron stream, the electrode acquires an electrostatic charge from the electrons which hit it. The amount of charge depends on the beam current, and focusing is automatic.

In practically all present-day television receivers, focusing is accomplished by changing the potential applied to a focusing anode built into the electron gun.

Beam Deflection

Now that we have produced and accelerated the beam and can manually control its intensity and focus, the beam must be given a horizontal and vertical movement within the area of the fluores-

cent screen. Two sets of deflecting plates with horizontal and vertical orientation (Fig. 1-9) are mounted in the neck of the tube. They are so arranged that the electron beam passes between each set of plates after it has sped through the anodes toward the screen. The complete assembly of a cathode-ray tube with electrostatic deflection is illustrated in Fig. 1-10.

Since the electrons in the beam are negatively charged, their movement is governed by the basic law of attraction and repulsion—"like charges repel one another, unlike charges attract one another." Therefore, a positively charged plate will attract electrons, and a negatively charged plate will repel them.

An electrostatic field exists between two adjacent plates of opposite polarity. When an electron is shot into an electrostatic field whose lines of force cross its path (Fig. 1-9), the electron tends to drift from its normal course toward the positively charged plate. The electron actually crosses the lines of force because of its own momentum, although the static lines pull the electron in their direction. The high speed at which the electron beam passes through the static field delays its de-

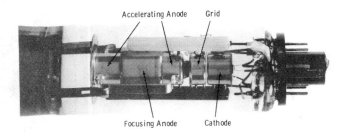

Fig. 1-8. Electron gun used in a modern, low-voltage, electrostatic-focus picture tube.

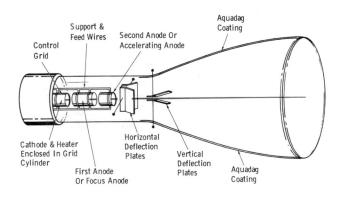

Fig. 1-10. Electrostatic focus and deflection.

flection slightly, thus preventing it from hitting the positive plate. Therefore, the amount the beam is deflected from its normal course depends on the velocity of the beam and the strength of the deflecting field. The horizontal and vertical deflections could be increased by increasing the distance between the point of deflection and the screen; this, of course, would lengthen the cathode-ray tube. The distance the beam or spot is moved across the screen by an applied voltage of one volt across the deflection plates is called the deflection sensitivity. In some specifications, the sensi-

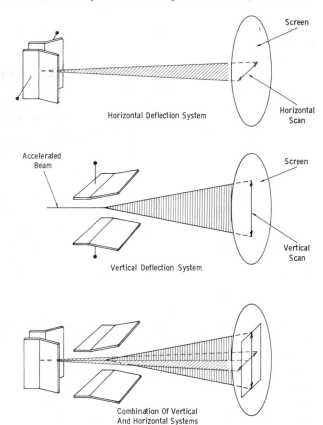

Fig. 1-9. Electrostatic beam-deflection system.

tivity of the horizontal-deflection plates will be greater than that of the vertical plates, since the horizontal plates are farther from the screen. However, in a cathode-ray tube, the velocity of the beam and the position of the electrodes are fixed; therefore, to increase the deflection, the deflecting voltage must be increased.

Another way of increasing the deflection sensitivity is by lengthening the deflection plates so that the static field is active on the beam for a longer period of time. In this case, the ends of the plates must be bent to form a flare (Fig. 1-9).

In early cathode-ray tubes, deflection voltage was obtained from single-ended amplifiers. One plate of each set of deflection plates was connected to the accelerating anode. When a voltage was applied to the deflection plates, a difference of potential existed between the accelerating anode and the deflection plates, causing defocus and a change in the beam velocity. This effect is called *astigmatism.*

In present-day cathode-ray tubes using electrostatic deflection, a separate terminal is provided for each deflection plate, making possible the use of push-pull deflection amplifiers. The average potential remains constant between the plates of either pair, because the increase in potential of one plate is equal to the decrease in potential of the other plate. Thus, any defocus or any change in beam velocity is minimized. Some tubes also have a ring or element between the horizontal- and vertical-deflection plates. This element is connected to the accelerating anode and prevents defocus due to any disturbing field between the pairs of deflecting plates.

The accelerating anode in modern cathode-ray equipment is connected to a variable voltage divider, usually called the astigmatism control. This control is adjusted for maximum roundness of the spot on the screen.

The electrostatic deflection systems just discussed are no longer used in television receivers. However, cathode-ray tubes employing electrostatic deflection are still used extensively in oscilloscopes. Almost all television receivers manufactured since 1950 have featured electromagnetic deflection, which will be discussed thoroughly in Chapter 2.

BEAM EFFECT ON THE FLUORESCENT SCREEN

The human eye can retain an image about $\frac{1}{16}$ of a second after the image disappears. This phenomenon is used in motion pictures, where a series of still pictures is projected on the screen so rapidly that the eye does not detect them as separate pictures. In a cathode-ray tube, the beam is swept so fast that the moving spot looks like a straight line. If the beam is swept over the same line or path at least 16 times a second, the spot looks to the viewer like a continuous line of light without flicker. Therefore, if the combined action of the horizontal- and vertical-deflection voltages sweeps the beam horizontally and vertically at the same time, a frame of light will appear on the screen (Fig. 1-9). In other words, a small spot of light appears at the point where the electron beam strikes the screen, and then, if the beam is deflected left to right and top to bottom very rapidly, the whole screen is illuminated. This frame of light, the intensity of which can be controlled, is called a raster when used in television. Since the raster consists of small spots of light, a signal that will modulate the beam and cause each spot to vary in brilliancy can be inserted into the control-grid circuit of the cathode-ray tube. In this way, a picture is formed.

QUESTIONS

1. Name the main sections of the electron gun in a cathode-ray tube.

2. What is the electron source in modern cathode-ray tubes?

3. What determines the intensity of light on the screen of the picture tube?

4. Which anode in the cathode-ray tube is used to narrow the beam?

5. What term is used to refer to the distance the beam is moved across the screen by a voltage of one volt across the deflection plates?

6. The amount the beam is deflected off its normal course depends on what two factors?

7. What is the name of the frame of light that appears on a television cathode-ray tube when the whole screen is illuminated?

EXERCISES

1. Sketch the internal construction of the electrostatic cathode-ray tube and identify each element.

2. Give the function of each element in Exercise 1.

CHAPTER 2

Cathode-Ray Tube—Electromagnetic Control of the Beam

Up to this point, our discussion of beam formation and control has been primarily about cathode-ray tubes using electrostatic control methods. A second type of cathode-ray electron-beam control is obtained through electromagnetic deflection by varying the relative force, position, or area of magnetic fields adjacent to the beam. This method of beam control is used exclusively in present-day television receivers.

The construction of a typical electromagnetically controlled cathode-ray tube is shown in Fig. 2-1. Notice the similarity between the electron gun in this tube and the electrostatic type shown previously.

In earlier television receivers, focusing was accomplished by an adjustable magnet placed externally around the neck of the picture tube. As mentioned before, focusing is currently accomplished by changing the dc potential on a focusing anode built into the electron gun. Even though external magnetic focusing devices are no longer used with present-day receivers, a discussion of their principles will be helpful in understanding the newer method.

Fig. 2-2 shows an early electromagnetically controlled cathode-ray tube using an external focusing device. To more easily understand the overall operation of this tube, let us review the effect of magnetic fields on an electron beam.

EFFECT OF MAGNETIC FIELDS ON AN ELECTRON BEAM

The stream of electrons from the beam source may be considered equivalent to a stream of electrons in a solid conductor carrying direct current. The effect of an external magnetic field on either stream will be the same, since any flow of electrons produces its own magnetic field. The direction of the electron flow and the magnetic lines it produces are at right angles (Fig. 2-3).

If this current-carrying conductor is placed in an existing magnetic field with the conductor parallel to the lines of force of this field, no force will be exerted on the electron stream. The magnetic lines from the two sources are at right angles, neither aiding nor opposing one another; therefore, no interaction will result (Fig. 2-4).

However, if the conductor is at right angles to the existing field (Fig. 2-5), a torque or distortion of the magnetic lines will tend to move the conductor from the field. This is because the lines of the two fields are opposing on one side of the current-carrying conductor or stream of electrons, and aiding on the other side.

Fig. 2-5 shows the electron stream or conductor at a right angle to the external magnetic field. However, an electron stream entering the magnetic field at any angle other than parallel will also be affected by the external magnetic field. This effect will be proportionate to the amount of angular variation.

Let us apply the foregoing in terms of electromagnetic control of the beam in a cathode-ray tube.

BEAM FOCUSING

Note in Fig. 2-2 that the focusing device was placed along the neck of the tube. Since its mag-

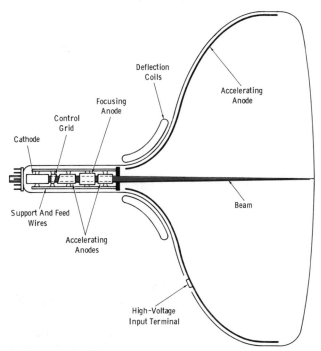

Fig. 2-1. Typical modern cathode-ray tube which uses external electromagnetic deflection and internal electrostatic focusing.

netic field controlled the size of the electron beam and caused the formation of a small spot of light on the tube face, no internal focusing elements were required.

Provision was made for moving the focus coil along the neck of the tube. With the coil in approximately the correct position, control of fine focus was obtained by varying the direct current through the coil. The control used for this purpose was known as the focus control and normally consisted of a potentiometer. The principle of the focus coil is shown in Fig. 2-6.

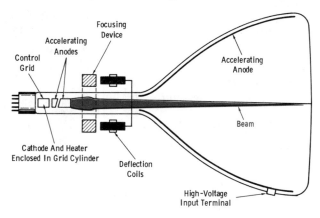

Fig. 2-2. Early cathode-ray tube which used external electromagnetic deflection and external magnetic focusing.

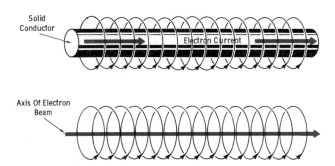

Fig. 2-3. Similarity of a solid conductor and an electron beam.

The first focusing devices were electromagnets. These electromagnets, wound with many turns of wire, were placed on the neck of the cathode-ray tube to concentrate the magnetic field there. Thus, the beam of electrons inside the tube was sur-

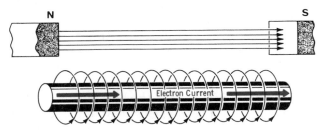

Fig. 2-4. Conductor parallel to a magnetic field.

rounded with parallel lines of magnetic force. By concentrating these lines of force, two results were obtained: (1) Less magnetic strength was necessary than with other structures. (2) Stray fields were lessened; therefore, they were less

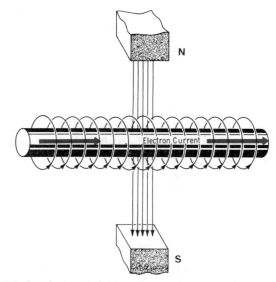

Fig. 2-5. Conductor at right angles to a magnetic field.

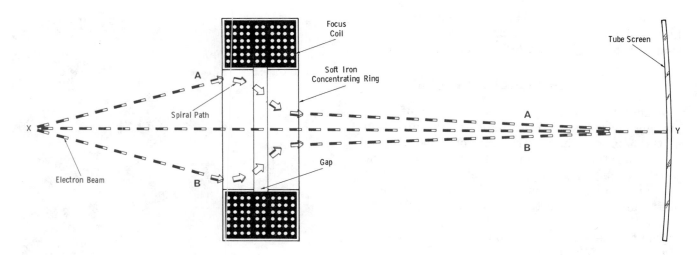

Fig. 2-6. Action of focus coil on electron beam.

likely to affect the action of the other beam-control elements.

To generate a magnetic field, the focus coil had to be supplied with a well-filtered direct current. For this reason, the focus coil was soon replaced by a permanent-magnet (pm) focusing device. Coarse focusing was obtained by proper placement of this device along the tube neck. For fine focusing, a fine-focus screw moved a soft-iron magnetic shunt back and forth and thus varied the magnetic strength. The pm focus device acted the same as the focus coil pictured in Fig. 2-6.

Referring to Fig. 2-4, we can see that any electrons traveling along a path parallel to the lines of an external magnetic field will not be affected by that field; they will continue to travel in a straight line. Axis XY of Fig. 2-6 shows this path.

Referring to Fig. 2-5, we can see that a stream of electrons entering an external magnetic field at right angles will be deflected out of the field. Any time this stream of electrons enters the field at an angle other than parallel, the beam will be deflected somewhat. Lines A and B of Fig. 2-6 represent the paths of electrons not parallel to axis XY. Therefore, since this beam is entering the magnetic field of the focus coil at an angle, the stream will be pushed sideways. Because the electrons are traveling very rapidly and each electron has its own magnetic field, the beam within the focus field will follow a path similar to the thread of a wood screw. With a proper balance of the beam velocity, the magnetic field produced by the focusing device, and the potential applied to the accelerating anode, the electron beam will leave the focus field in a converging stream, having its focal point at the fluorescent screen.

In present-day television receivers, the once-popular external magnetic focusing device has been replaced by a focusing anode built into the electron gun of the cathode-ray tube (Fig. 2-1). Focusing is accomplished by changing the voltage applied to the focus anode. The voltage applied to the focus anode is changed either by selecting the voltage tap that provides the best focus (Fig. 2-7A) or by means of a continuously variable focus control (Fig. 2-7B). The focusing action obtained by varying the potential on the focusing anode is similar to that shown previously in Fig. 1-7.

Very few modern black-and-white television receivers have a focus control as shown in Fig. 2-7B. Most sets use voltage taps as the method of vary-

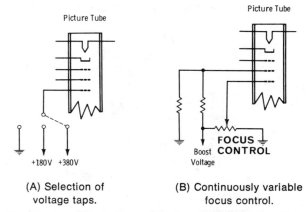

(A) Selection of voltage taps.

(B) Continuously variable focus control.

Fig. 2-7. Common methods of varying the focus voltage in present-day receivers.

ing the focus voltage. Some inexpensive portables have a fixed focus voltage, and there is no provision for any adjustment of the focus.

BEAM DEFLECTION

A magnetic field will deflect a beam of fast-moving electrons at right angles to the direction of the field and the electron motion. (See Fig. 2-8A). Therefore, electromagnetic deflection of the beam may be obtained by two sets of coils. These coils are arranged horizontally and vertically over the neck of the tube, near the bulge of the bulb (Fig. 2-1).

The two sets of coils are mounted in an assembly called a deflection yoke that has four windings, two for horizontal deflection and two for vertical deflection. A typical deflection yoke is shown in Fig. 2-9. Fig. 2-10 shows the coil construction and assembly of a complete deflection yoke. The two horizontal coils are opposite each other and connected in series for correct polarity. Thus, the magnetic field passes through the neck of the tube at right angles to the path of the beam and is oriented vertically for horizontal deflection.

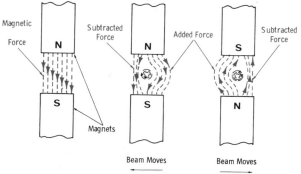

(A) Movement of electron beam in deflection fields.

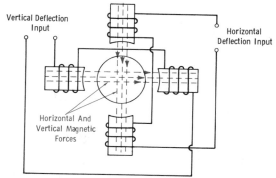

(B) Arrangement of magnetic coils in deflection yoke.

Fig. 2-8. Action and position of deflection coils.

Fig. 2-9. Typical magnetic deflection yoke.

The vertical-deflection coils are arranged and connected in the same manner, but oriented horizontally for vertical deflection (Fig. 2-8B).

Due to the arrangement of the horizontal and vertical deflection coils, the spot produced by the beam can be moved anywhere on the screen by passing the correct amount of current through each set of coils. If a sawtooth of current is passed through the horizontal coils, it will cause the spot to move from left to right across the screen and then fly back. Similarly, if a sawtooth of current is passed through the vertical coils, the beam will be made to move from top to bottom and fly back. The combined action of the horizontal and vertical fields will produce a frame of light, or raster.

A complete discussion of the development of these sawtooth waveforms will be given in the section covering horizontal and vertical oscillators.

BEAM CENTERING

The beam in an electromagnetically controlled cathode-ray tube can be centered in several ways. One of the early methods was a potentiometer which varied the amount of direct current flowing through the deflection-yoke windings. A direct current provides a fixed magnetic bias for positioning the beam at a point which becomes the center of deflection after deflection voltages are applied to the yoke. This method is still used in some color sets.

Centering-control potentiometers were soon abandoned in favor of a method that allowed cir-

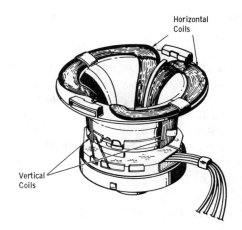

(A) Complete assembly.

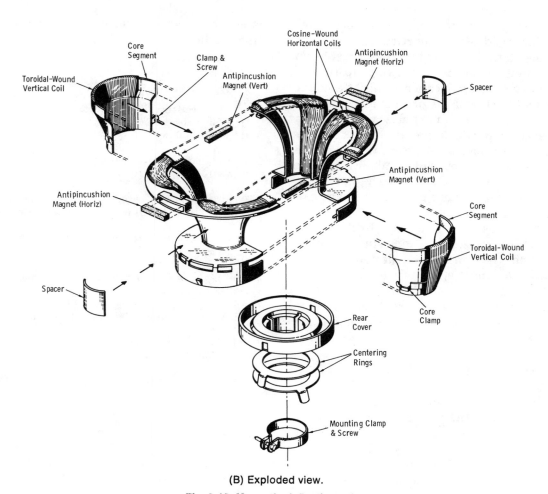

(B) Exploded view.

Fig. 2-10. Magnetic deflection yoke.

cuit simplification. Because the magnetic field of a focusing device exerts a magnetic bias on the electron beam, the position as well as the focus of the beam can be controlled. The mounting of the focusing device was made variable for variable bias. For further simplification, the focusing-device mounting was made stationary, and a movable magnetic shunt was added. The shunt concentrated the magnetic field, and as the shunt was moved, it centered the electron beam. The focus was adjusted by moving the shunt along the axis of the tube.

With the development of electrostatic focusing, it became necessary to use a separate device for centering. The present method of centering involves the use of two magnetized rings mounted on the neck of the tube directly behind the deflection yoke. Centering is accomplished by rotating the rings separately or simultaneously in the appropriate direction. The location of the centering rings is shown in Fig. 2-9.

REMOVAL OF IONS FROM ELECTRON BEAM

The emitted electrons from the cathode are mixed with charged atomic particles called ions. These ions are in the tube for two reasons: (1) no matter how well the elements making up the internal tube structure are cleaned, a slight amount of foreign material will be present; (2) as the cathode is heated, small particles will tend to break loose from it. Each ion is approximately 2000 times heavier than an electron. If ions strike the screen, a brown spot will gradually appear because ions remove the phosphor. The net result is a spot on the screen where no picture image can be produced.

The early electrostatically deflected tubes were not affected by the presence of ions since the electron beam and the ions are deflected simultaneously. However, magnetic fields do not greatly deflect an ion, and tubes using magnetic deflection systems must have some provision for preventing a concentration of ions from reaching the center of the phosphor screen; otherwise, a spot will result.

Two methods have been used to remove the ions. The early method involved the use of the "bent-gun" tube. In this type of tube, part of the electron gun is at an angle to the rest of the gun. As the ions and electrons emerge from the first part of the gun, they would normally be eliminated by striking the other part of the gun. A single magnet at the proper point on the neck of the tube bends the electrons back to the center of the second part of the gun, and they continue on to strike the screen.

The present method of removing ions uses an extremely thin film of aluminum on the beam side of the phosphor screen. The aluminum is thin enough that the electrons can pass through and strike the phosphor. Since the ions are larger, they will not penetrate the aluminum and, therefore, will not strike the phosphor. The aluminum film also provides better contrast and more brilliance of the picture. Aluminized picture tubes employ a straight gun and do not require an ion trap.

The general types of picture tubes and their requirements for deflection have been covered. Before discussing the method of translating received signals into visual patterns on the face of the picture tube, let us see how the televised scene is converted into a video signal at the transmitting studio.

QUESTIONS

1. Is the direction of the magnetic lines produced by electron flow *parallel* or at a *right angle* to the direction of the electron flow?

2. Under what conditions will no force be exerted on an electron stream by a surrounding magnetic field?

3. What element is now employed in the cathode-ray tube in lieu of an external magnetic focusing device?

4. How is focusing normally accomplished in modern television receivers?

5. How is the beam deflected in an electromagnetically controlled tube?

6. What type of current is passed through the yoke?

7. How is beam centering normally accomplished in modern television receivers?

8. What portion of the beam must be removed before the beam strikes the phosphor screen? Why?

EXERCISES

1. Sketch the electromagnetically controlled tube and show its internal elements and external components. Identify each.

2. Give the function of each element and component shown in Exercise 1.

CHAPTER 3

The Camera Tube

A brief description of the internal construction and operation of camera tubes will be helpful in associating the scanning technique and transmitted picture with that of the receiving picture tube. The camera tubes discussed in this chapter are the image orthicon and the vidicon. These tubes are used in both black-and-white cameras and color cameras. The monoscope, which is used to produce television test patterns, will also be described briefly.

IMAGE ORTHICON

The image orthicon (Fig. 3-1) is intended for use in both black-and-white and color cameras for outdoor and studio pickup. It is very stable in performance at all incident light levels. For a better understanding of the operation of the image orthicon, refer to Fig. 3-2 while studying the following paragraphs.

Light from the scene being televised is focused on the semitransparent photocathode. This photocathode emits electrons proportional to the amount of light striking the area. These electrons are accelerated toward the target by grid 6 and focused by the magnetic field produced by an external focus coil. The target consists of a special thin glass disc with a fine mesh screen on the photocathode side. Focusing is also accomplished by varying the potential on the semitransparent photocathode.

When the electrons strike the target, secondary emission from the glass disc takes place. These secondary electrons are collected by the wire mesh, which is maintained at a constant potential of approximately one volt. This limits the potential of the glass disc and accounts for its stability in changing light intensities. As electrons are emitted from the photocathode side of the glass disc, positive charges are built up on the other side. These charges vary with the amount of electrons emitted. Thus, a pattern of positive charges corresponding to the light intensities of the scene being televised is set up. This constitutes the image section of the image orthicon. The action described is completely independent from the electron beam and scanning circuits of the tube.

The backside of the target is scanned with a low-velocity beam from the electron gun. The beam is focused by the magnetic field generated by an external coil and by the electrostatic field of grid 4. The potential applied to grid 5 adjusts the decelerating field between grid 4 and the target. As the low-velocity beam strikes the target, it is turned back and focused on dynode 1, the first element of an electron multiplier. However, when the beam is turned back from the target, some electrons are taken from the beam to neutralize the charge on the glass. The greater the charge on the glass, the more electrons are taken from the beam. Thus, when the beam scans a more positively charged area, corresponding to an area of brighter light intensity, fewer electrons are returned to dynode 1. This action leaves the scanned side of the target negatively charged, while the opposite side is positively charged. Because the glass-disc target is extremely thin, these charges neutralize themselves by conduction through the glass. This neutralization takes place in less than the time of one frame.

As the amplitude-modulated stream of electrons strikes dynode 1, secondary electrons are emitted. Several secondary electrons are emitted for each primary electron striking the element. These free

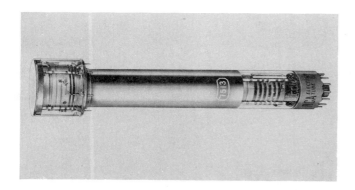

Fig. 3-1. Image orthicon.

electrons are then accelerated toward dynode 2. As the electrons strike the element, more secondary emission takes place at dynode 2. This same process continues on through dynodes 3, 4, and 5. The electrons are finally collected by the anode. Thus, the electrons returned to dynode 1 are amplified, or multiplied, many times before the signal reaches the anode. The multiplication per element equals the difference between secondary electrons emitted from the element and electrons striking the element. The approximate gain of the multiplier section of this tube is 500. The load resistor for the image orthicon is connected from the anode to the power supply. More current in the multiplier, which corresponds to a dark area in the television scene, causes more current in the load, giving a negative output. A brighter area causes less current, providing a less-negative output. Since the polarity of the television signal is always given in reference to the black portion of picture, the output of the image orthicon will be of negative polarity.

VIDICON

The vidicon (Fig. 3-3) is a small camera tube suitable for either broadcasting or closed-circuit applications. The vidicon is used in both black-and-white and color television cameras.

The structural arrangement of the vidicon, shown in Fig. 3-4, consists of a target, a fine mesh screen (grid 4), a beam-focusing electrode (grid 3), and an electron gun. The target, or light-sensitive element, of the vidicon consists of a transparent conducting film on the inner surface of the glass faceplate and a thin layer of photo-conductive material on the scanning side of the film. Grid 4, the wire-mesh screen, is adjacent to the photoconductive layer, and grid 3 is connected to grid 4. The photoconductive layer is an insula-

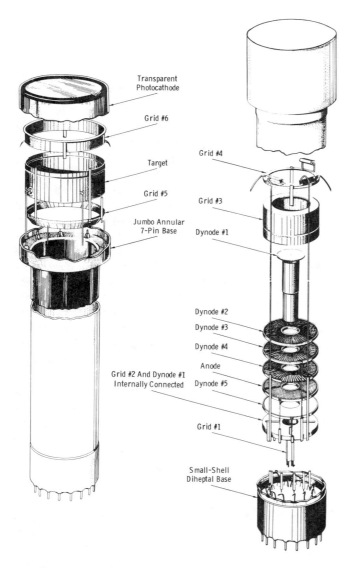

Fig. 3-2. Exploded view showing the construction of an image orthicon.

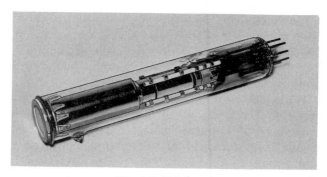

Fig. 3-3. Vidicon.

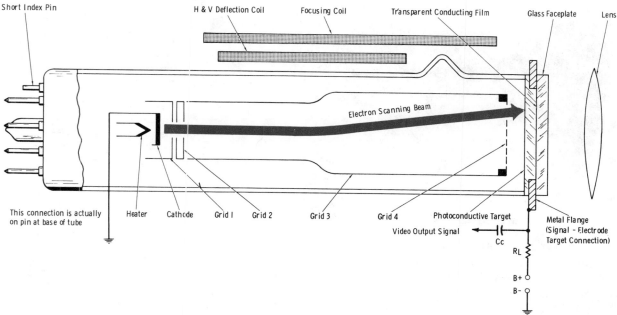

Fig. 3-4. Structural arrangement of the vidicon.

tor when there is no light on the faceplate. When an image is focused on the faceplate, the illuminated areas of the photoconductive area will become slightly conductive.

The scanning side of the photoconductive layer is scanned by a low-velocity beam produced by the electron gun. The electron gun consists of a cathode, a control grid (grid 1), and the accelerating grid (grid 2). The electron beam is focused by the magnetic field of the focusing coil and the electrostatic field produced by grids 3 and 4. Grid 4 provides a uniform decelerating field between itself and the photoconductive layer. This allows the scanning beam to strike the photoconductive layer perpendicularly, which is a necessary condition for linear scanning.

The target is connected electrically to a metal ring around the front of the tube which serves as the signal electrode. The load resistor, R_L, is connected between the signal electrode and the B+ power supply. The complete circuit for the vidicon consists of the scanning beam, the photoconductive layer, the load resistor, and the power supply, as shown in Fig. 3-4.

When there is no light on the faceplate of the vidicon, the photoconductive layer is an insulator with a very high resistance. Therefore, the two surfaces of the target may be considered as two plates forming a capacitor with a dielectric resistance inversely proportional to the amount of light

on the photoconductive material. The faceplate side of the target is connected to the positive potential of the signal electrode, and the scanning side is essentially floating.

When the gun side of the photoconductive material, with its positive-potential pattern, is scanned by the electron beam, electrons are deposited from the beam until the potential of the scanned surface of the photoconductive layer is reduced to the potential of the cathode. The remaining electrons are turned back to form a return beam, which is not used. Although a considerable potential exists between the two surfaces of the target, there is very little current flow due to the high resistance of the dielectric when there is no light on the photoconductive layer. However, when any portion of the photoconductive material is illuminated, the resistance of that portion decreases and there is some conduction. This causes the corresponding area on the gun side of the photoconductive material to increase its positive potential, and, therefore, more electrons from the electron beam are deposited on the target. The increased beam current through the load resistor produces a greater negative voltage drop, which constitutes the video signal. Since the black portion of the picture produces a less-negative output, the vidicon is said to have a positive output polarity. This is opposite to the negative output of the image orthicon discussed earlier.

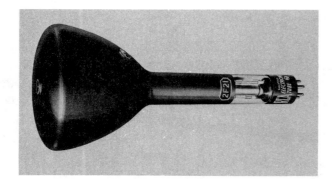

Fig. 3-5 Monoscope.

MONOSCOPE

Another type of cathode-ray tube used in the development of television signals is the monoscope, shown in Fig. 3-5. The monoscope provides a test pattern which is used for testing and adjusting studio equipment. When transmitted by the station, this pattern is also useful for proper adjustment of receiving equipment. The primary difference between this tube and the other camera tubes discussed previously is that it has a test pattern inscribed inside the face of the tube. This test pattern is reproduced as the video signal.

The difference in the amount of secondary emission of electrons between two different materials is used to produce the output from a monoscope. Usually, a sheet of aluminum, which has high emission, is marked with a high-carbon-content ink. Carbon has fairly low emission, and, as the electron beam scans the entire pattern, secondary electrons are emitted from both materials in proportion to their emission ratios. Any pattern, with any line shape, may be drawn on the aluminum sheet. The monoscope is a stable video-signal source and provides both the television broadcast engineer and the service technician with a useful test pattern.

SCANNING

The three picture-generating tubes discussed here have associated external focus and deflection elements. These focusing and deflection elements cause the electron beam to scan the image surface, or target, at the front of the tube.

The video signal carries the picture information to be transmitted over the air. Since the timing of the scanning is important to properly reproduce the picture, the video signal must contain other

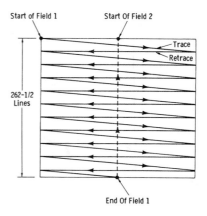

(A) Field 1.

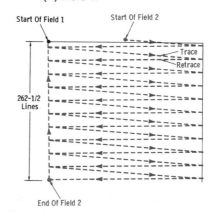

(B) Field 2.

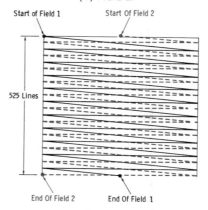

(C) Complete frame.

(D) Sawtooth deflection currents.

Fig. 3-6. Principles of interlaced scanning.

information in the form of electrical pulses. One of these is termed a blanking pulse, and its purpose is to blank out the cathode-ray beam in the picture tube during retrace time. Other pulses, the synchronizing pulses, are used by the receiver to synchronize the horizontal and vertical sawtooth generators. The path traveled by the beam across the screen of the picture tube should be identical to the path traveled by the beam in the camera tube, so that the picture may be reconstructed in the correct sequence at the receiver.

For picture resolution, the U.S. standards for television broadcasts are 30 frames per second, each frame having 525 lines with interlaced scanning. (If alternate lines are transmitted so that two series of lines are necessary to produce a complete frame, the system is called interlacing.)

Therefore, to produce one frame of 525 interlaced lines, $262\frac{1}{2}$ lines are scanned on the first down sweep of vertical deflection, and then the beam returns to the top and scans $262\frac{1}{2}$ alternate lines. The horizontal and vertical scanning traces are the result of a sawtooth current passing through the respective deflection coils. The rapid return of the electron beam, or retrace, for the start of the succeeding scanning function is a result of the rapidly decreasing portion of the sawtooth current. To produce interlaced scanning with 30 complete frames per second having 525 lines each, the vertical-sweep frequency must be 60 Hz, and the horizontal-sweep frequency must be 15,750 Hz. Fig. 3-6 further explains the complete scanning operation.

QUESTIONS

1. What is the purpose of the camera tube?

2. What is the scene focused on in an image orthicon?

3. Which of the following tubes is used to produce a test pattern for adjusting and testing studio and receiving equipment?
   (1) Vidicon.
   (2) Monoscope.
   (3) Image orthicon.

4. How many frames per second and how many horizontal lines per second make up the television picture in the U.S.?

5. What are the vertical- and horizontal-sweep frequencies?

6. What method of scanning is employed in television broadcasting?

7. What is the output polarity of the vidicon?

8. What is the function of the target in the image orthicon?

9. In the vidicon, what is the function of the photoconductive material?

10. What information other than the picture information must be contained in the video signal?

CHAPTER 4

Power Supplies

Television receivers require at least two separate and completely different types of supplies, one capable of supplying low voltage and high current, and another for producing high voltage and low current. The low-voltage power supply is used to power the oscillators, amplifiers, and similar circuits where potentials of 300 volts or less are required. The high-voltage supply, however, must be capable of producing anywhere from 7000 to 30,000 volts to be applied to the accelerating anode of the cathode-ray tube. Since high-voltage power supplies are part of the horizontal deflection system in modern television receivers, they will be covered in Chapter 8.

A number of different circuit arrangements have been employed in past years to produce the required low and high voltages. Earlier sets employed large transformers and rectifier tubes to meet the low-voltage, high-current requirement, while some employed an rf high-voltage circuit to provide high voltages. In time, the power supply circuits were improved and became more simplified and compact. Rectifier tubes, for example, were gradually replaced by silicon rectifying devices in the low-voltage circuits. During this time, vacuum tubes for other circuits were improved, and, as a result, lower dc supply voltages were required for operation. Many set manufacturers began to eliminate the bulky low-voltage transformer and employed a power supply that operated directly from the 117-volt ac power line. Power supplies of this type require that the tube filaments be connected in a series arrangement whereby the applied voltage is divided among them. This type of power supply is still a popular one in many present-day television receivers, and solid-state rectifiers (usually silicon) are being used instead of tubes.

In this chapter we will discuss not only the latest types of low-voltage power supply circuits but also some of the earlier methods used to obtain these voltages and currents. This is to show why some of these changes were made and to provide "building blocks" for an easier understanding of television power-supply circuits.

The signal-processing portion of the television receiver presents a power requirement not greatly different in voltage range from that of other electronic devices. Therefore, this portion of the power supply is similar to that found in large radios or stereo amplifiers. In general, the voltage requirement is no more than 300 volts in modern tube-type receivers, and may be as low as 12 volts in the solid-state sets.

The power supplies used in television receivers require good regulation and filtering to operate the sawtooth oscillators for deflecting the electron beam in the cathode-ray tube. The oscillators in the deflection circuits tend to produce currents in the power supply which, if not properly filtered, would appear as serious hum modulation in both the sound and the picture.

Early television receivers required heavy-duty power supplies to supply the current drawn by the 30 to 40 tubes used in the sets. These supplies employed from one to three rectifier tubes, depending on the current requirements and the designer's preference as to voltage and current distribution. Different voltages—both positive and negative, in some instances—were needed by the other stages in the receiver, and an elaborate voltage-divider or distribution network was needed to produce these voltages. In early receivers, the dc current drawn by several stages was passed through the electromagnetic focus coil, which did double duty as a filter choke.

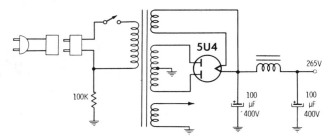

Fig. 4-1. A power supply using a power transformer and a
vacuum-tube rectifier.

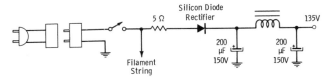

Fig. 4-2. Schematic of a power supply using a
semiconductor rectifier.

As the receivers were simplified, the voltages and currents needed decreased, and the power supply was also simplified. A simplified power supply using a single full-wave rectifier tube is shown in Fig. 4-1. Only one output voltage is directly available in this configuration. Stages requiring lower voltages are provided with individual voltage-dropping resistors. This type of supply, with many minor modifications and additions, was used in numerous television receivers.

SOLID-STATE RECTIFIERS

The introduction of semiconductor rectifiers offered an opportunity to simplify and lighten receivers by eliminating the heavy and bulky power transformer. This is possible because semiconductor rectifiers do not require the high-current filament winding. The remaining tubes in the receiver usually have their filaments connected in series, although some sets have employed a separate filament transformer to supply the tube heaters.

The first semiconductor rectifiers consisted of a series of aluminum plates coated on one side with selenium. These selenium rectifiers were eventually replaced with silicon diodes, which are much smaller in size and have considerably higher current ratings. Silicon rectifiers slightly larger than a match head are available with current ratings of 2 amperes or more. A typical half-wave

rectifier circuit using a silicon diode is shown in Fig. 4-2. Since only 135 volts is available from the circuit shown in Fig. 4-2, the other circuits in the receiver must be designed to operate at this rather low voltage. Those few circuits that require more voltage derive it from the boosted B+ voltage produced by the damper circuit. The development of the boost voltage will be discussed in Chapter 8.

The circuit shown in Fig. 4-3 was used in many tube-type television receivers, including several color sets. This doubler configuration produces a voltage slightly higher than twice the line voltage. The operation of the doubler is relatively simple. A dc voltage of about 145 volts is developed across C1 due to the conduction of X2 during the negative half cycles. When diode X1 conducts during the positive half cycles, the voltage developed across C1 is in series with the line voltage. Therefore, the voltage produced across C2 will be approximately 290 volts. It is easy to see that the voltage rating of C2 must be twice that of C1. The 4.7-ohm resistor in series with the line acts as both a fuse and a surge resistor. The surge resistor is necessary to protect the rectifiers during the initial charging of the electrolytic capacitors. Without the resistor, the surge caused by this initial charging would exceed the voltage ratings of the rectifiers and damage them.

The circuits shown in Figs. 4-2 and 4-3 have one disadvantage: the chassis is connected to one side of the power line. The user could receive a dangerous electrical shock between the chassis and

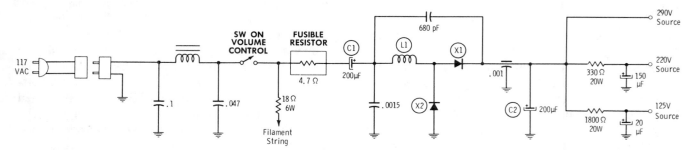

Fig. 4-3. Schematic of a power supply using a voltage-doubler circuit.

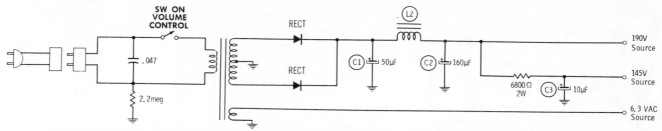

Fig. 4-4. Schematic of a power supply using a transformer and solid-state rectifiers in a full-wave configuration.

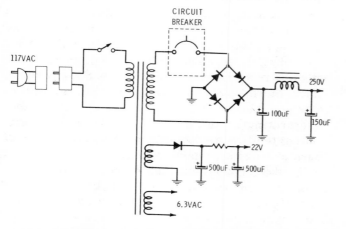

Fig. 4-5. Power supply employed in a television receiver using both tubes and transistors.

any grounded metal object if the ac plug is inserted so that it connects the high (or "hot") side of the line to the chassis. A power transformer can be used with solid-state rectifiers, as shown in the full-wave circuit of Fig. 4-4, to eliminate this shock hazard.

Some television receivers use a combination of tubes and solid-state devices. These combination

sets (usually referred to as hybrids) generally employ a power supply similar to that shown in Fig. 4-4, but with an additional low-voltage winding on the power transformer. The purpose of this extra winding is to provide low-voltage power for the solid-state devices. Such a power supply is shown in Fig. 4-5. Note the 22-volt source used to supply the solid-state devices.

REGULATION AND FILTERING

Fig. 4-6 shows a power supply for an all solid-state receiver. Generally, such power supplies are well regulated to provide a relatively constant, ripple-free output, since the operation of a transistor can be seriously affected by a small voltage change. The regulator circuit shown in Fig. 4-6 is typical of the power-supply regulators found in transistor television receivers. The regulator transistor shown in Fig. 4-6 is called a series regulator, since it is in series with the regulated portion of the load. The 12-volt source voltage at the collector of the regulator transistor also appears at the emitter of the error amplifier. Since the base of the error amplifier is held at a fixed voltage,

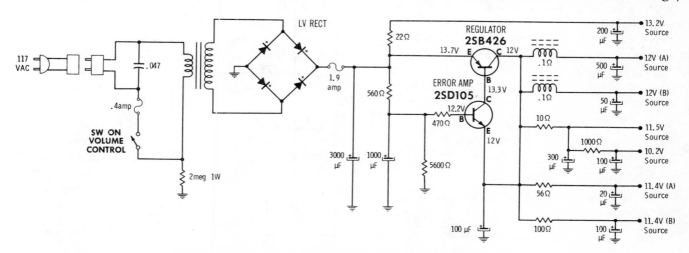

Fig. 4-6. Typical power supply used in a solid-state television receiver.

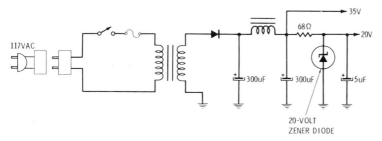

Fig. 4-7. A zener diode used as a voltage regulator.

any change of the emitter voltage will cause a corresponding change in the collector-to-emitter resistance of the error amplifier. This, in turn, will cause a change in the emitter-base (bias) current of the regulator transistor. The resulting change in the collector-to-emitter resistance of the regulator transistor will cause a change in the source voltage to oppose the voltage change which initiated the regulating action.

The action of the regulator circuit is instantaneous and will tend to filter out any ripple voltage appearing at the emitter of the error amplifier. For this reason, this type of circuit is often referred to as an electronic filter or an active power filter.

There are many variations of the circuit shown in Fig. 4-6, some using as many as four transistors. Often, a zener diode is used to establish a fixed voltage on the base or emitter of the error

amplifier. In other circuits, the regulator transistor is in parallel with the regulated load and is known as a shunt regulator. In some receivers, the only regulation may be a zener diode, which is in parallel with the load, and a dropping resistor, which is in series with the load. Such a circuit is shown in Fig. 4-7. The zener diode will prevent the voltage applied to the load from rising above the zener rating of the diode. However, if the load voltage tends to be lower than the zener voltage, there will be no regulating action.

Fig. 4-8 shows the regulated power supply for a solid-state television receiver that can be operated from 120 volts ac or a 12-volt dc battery pack. The switching from ac to dc operation is automatically made by switch S901 when the ac power cord is unplugged from the set. If the batteries are connected with reverse polarity during dc operation, capacitor C901 and the ICs in other cir-

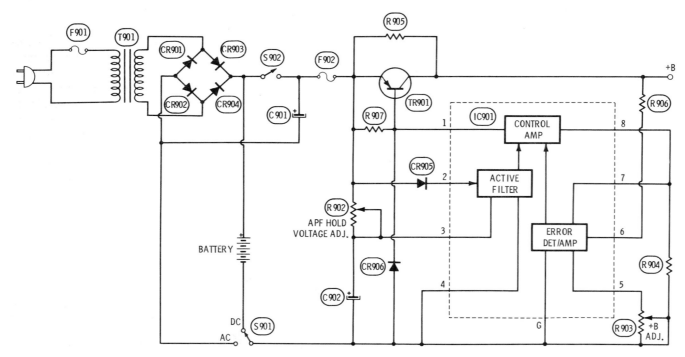

Fig. 4-8. Regulated power supply that can be operated from 120 volts ac or 12 volts dc.

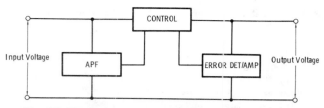

Fig. 4-9. Block diagram of voltage regulator and active power filter circuit.

cuits may be damaged. To prevent this, C901 is not connected in the circuit during dc operation. In addition, diodes CR905 and CR906 prevent damage to the other circuits if the battery is connected with reverse polarity.

The circuit consisting of constant-voltage control transistor TR901 and IC901 is a combination voltage regulator and active power filter shown by the block diagram in Fig. 4-9. The function performed by this circuit is determined by the level of the input voltage. When the input voltage is 12.4 volts or more, this circuit performs the regulating function that delivers the constant output voltage. If the output voltage changes for any reason, this change is detected and amplified by the error detector/amplifier. An error signal is then sent to the control section which reacts to return the output voltage to a predetermined level. The control section consists of the control amplifier and constant-voltage control transistor TR901.

When the input voltage falls below 12.4 volts, the circuit functions as an active power filter. The APF section in Fig. 4-9 maintains the difference between the input voltage and the output voltage at about 1 volt. At the same time, the output voltage is almost the same as the voltage at pin 3 of IC901. The voltage at pin 3 is sufficiently smoothed by filter capacitor C902 so that there is no objectionable ripple in the output voltage. The active power filter functions only when using a dc power source, and the voltage regulator functions only when the receiver is operated from an ac power source.

Since the supply voltages used with tubes are not nearly as critical as with transistors, voltage regulators are rarely found in the power supplies of tube-type sets. However, these tube-type receivers do require adequate filtering. Any 60-hertz line ripple present in the output of the power supply will show up as hum modulation in both the picture and the sound. As stated previously, the sweep oscillators used in the deflection circuits can cause the same problem, unless adequate de-

coupling is employed. Therefore, the power supply must be properly filtered.

The filtering components employed in the power supply in Fig. 4-4 are typical of those found in larger tube-type sets. Electrolytic capacitor C1 is the input filter, and C2 is the output filter. These capacitors and choke coil L2 form the filter network which effectively removes most of the ripple from the power-supply output. Output capacitor C2 also provides decoupling for the various stages supplied from the 190-volt source. Capacitor C3 provides decoupling for the 145-volt source. In most tube-type portable television receivers, filter choke L2 is replaced by a resistor. A resistor provides adequate filtering when the current drain is low. Since the low-voltage, solid-state receivers require significantly higher current from the power supply, a choke coil is almost always used for filtering.

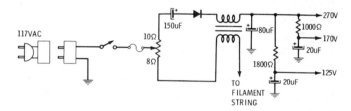

Fig. 4-10. Unique choke coil employing secondary winding to oppose ripple voltage.

The unique filter choke shown in Fig. 4-10 has been used in several imported tube-type receivers. A second winding is in series with the filament string and is phased in such a way that the voltage induced in the primary choke winding tends to cancel the ac ripple. If the connections to the secondary winding were reversed, the induced voltage would cause the ripple to increase. This would be due to the fact that the induced voltage would be aiding, rather than opposing, the ripple output from the power supply.

QUESTIONS

1. What two requirements make the power supplies in television receivers different from those used in large radios?

2. What is one advantage offered by a solid-state power supply connected in the voltage-doubler arrangement?

3. What is the disadvantage of a power-supply circuit that operates directly from the 117-volt ac power line and does not use a power transformer?

4. When different values of voltages are needed from the power supply, what are the methods used to obtain them?

5. What is the main difference in the requirements for a power supply used in a solid-state receiver compared to that used for a tube-type receiver?

6. Why is voltage regulation important in solid-state television receivers?

7. How does an electronic-filter circuit in a solid-state power supply operate?

8. Why must the power supply for a television receiver be well filtered?

EXERCISES

1. Show the following circuits used for the power supplies of present-day television receivers:
    (a) For a tube-type receiver.
    (b) For a solid-state receiver.

2. Draw the basic circuit for a voltage-doubler-type power supply.

Beam Deflection

CHAPTER 5

Resistance-Capacitance Circuit Characteristics

The elements in a cathode-ray tube provide an emitter, or source, of electrons; a means of forming an electron beam and accelerating it; and a phosphor-surfaced screen which will glow when bombarded by the stream of electrons. External deflection coils are also provided for moving the beam horizontally and vertically to form a frame of light, or raster, on the face of the picture tube.

The voltage or current waveforms required for deflecting the electron beam are obtained from sweep generators followed by special waveshaping circuits. The sweep generators are triggered by the synchronizing pulses derived from the transmitted signal. Thus, the sweep circuits of the receiver can be synchronized with those of the transmitter. The formation of a certain waveshape is required for a linear sweep. This waveshape is a complex voltage waveform used to obtain a sawtooth current in magnetic deflection coils. Some circuits will pass a waveform with negligible distortion. Other circuits distort greatly when generating, amplifying, or passing a waveform. The behavior of these distortion circuits can best be understood by studying a charging or discharging capacitor in series with a resistor.

Elementary theory states that a voltage, or IR drop, is developed across a resistor when electrons flow through it. The voltage developed by a current through a resistance is found by applying Ohm's law:

$$E = I \times R$$

where,
  E is in volts,
  I is in amperes,
  R is in ohms.

A further study of fundamentals reveals that a capacitor can store a charge of electrons. When charged, one plate contains more free electrons than the opposite plate. When the capacitor is completely discharged, both plates contain the same number of free electrons. When the accumulation of electrons on one plate exceeds the accumulation on the other plate, a potential difference exists across the terminals of the capacitor. This potential will continue to increase until it equals, for practical purposes, the applied or charging voltage. The value of the voltage developed by a charging capacitor is computed with the following equation:

$$E = \frac{Q}{C}$$

where,
  Q is in coulombs,
  C is in farads,
  E is in volts.

One coulomb is the quantity of electrons transferred when one ampere flows for one second.

RC CIRCUIT CHARGING

A capacitance and a resistance in a voltage-divider circuit (Fig. 5-1) develop a potential across their respective terminals. This circuit is commonly known as an RC circuit. Both Kirchhoff's and Ohm's laws apply to this circuit. Fig. 5-1 shows the voltage division across portion AB of the circuit diagram at various time positions on the graph after the switch is closed. As time progresses, voltage E_C on the capacitor gradually

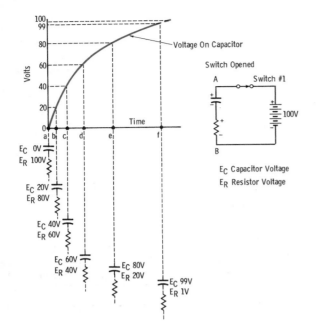

Fig. 5-1. Resistor-capacitor charging curve.

increases, while voltage E_R across the resistor gradually decreases.

When the switch is closed, electrons are displaced from the upper plate of the capacitor, and a positive charge is developed, causing electrons to be attracted to the lower plate through the resistor. The flow of electrons charges the capacitor. At the instant the electrons begin to flow, there is no charge on the capacitor, as seen at point *a* on the graph. Therefore, the applied voltage E across the divider must appear as a voltage drop across the resistor, and the initial charge current must equal E/R.

Kirchhoff's law states that the sum of the voltages in a closed circuit is equal to zero. Likewise, the sum of the voltage drops in a closed circuit must equal the applied voltage. Therefore, if 100 volts is applied to an RC circuit, 100 volts will appear across the resistor when the switch is closed. The graph shows that the instant the switch is closed, the entire applied voltage appears across resistor R, while the voltage across capacitor C is zero.

However, the current in the circuit soon charges the capacitor slightly, and a voltage appears across this capacitor. See position *b* of the voltage divider plotted on the graph. E_C is now 20 volts, and E_R is 80 volts, the sum of the two being equal to the applied voltage. As time elapses, E_C becomes greater and E_R smaller, as will be noted at

the time points *b*, *c*, *d*, *e*, and *f*. Actually, the capacitor voltage becomes a reactive voltage, or back pressure (opposite in polarity and opposed to the applied potential). This reactive voltage causes a decrease in the charging current and in the IR drop across the resistor; therefore, the capacitor charges at a slower rate.

This charging continues until the capacitor is almost fully charged. At this time the charging current and the voltage across R are practically zero. Theoretically, a capacitor never fully charges, and some minute voltage will always appear across the resistor. However, if the switch is closed long enough, an almost steady state is reached, and the capacitor is considered fully charged for all practical purposes.

RC CIRCUIT DISCHARGE

Suppose, at the time of point *f* on the charging curve (Fig. 5-1), the charging switch (No. 1) is thrown open and a discharging switch (No. 2) is closed, as shown in Fig. 5-2. Note that the capacitor voltage reaches a value of 99 volts. This value would have been slightly higher if the charging circuit had been left closed longer.

In Fig. 5-2, the battery switch is open. A short-circuit path is switched across the divider. The 99 volts of potential stored by the capacitor now be-

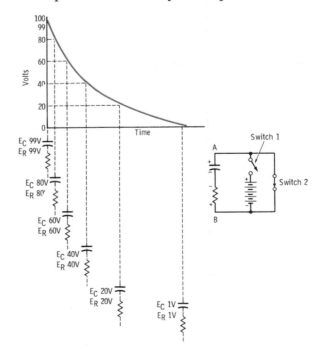

Fig. 5-2. Resistor-capacitor discharging curve.

comes the applied voltage of the discharge circuit, and electrons flow around the circuit. The discharge current is in a direction opposite to the charging current, and an IR drop will develop across the resistor. The voltage drop across the resistor due to the discharge current will be opposite in polarity from the voltage drop developed by the charging current.

During discharge, the capacitor voltage drops from its initial value and, representing the applied voltage of the discharge circuit, will equal the voltage drop across the resistor (Kirchhoff's law). Since the capacitor voltage now represents the applied voltage of the discharge circuit, E_C and E_R will slowly approach zero together.

The charging curve is not linear throughout. However, the charge portion of the curve (Fig. 5-1) up to 40 volts is practically straight. Because of this linearity, we will be more concerned with this portion of the curve in sweep circuits. This is the most important point of the discussion and should be kept in mind for future reference. Also note that the capacitor voltage does not reverse in polarity during the charge and discharge cycle. This is not true with the resistor voltage, because the current through the resistor actually reverses its direction between the charge and discharge period.

TIME CONSTANTS OF AN RC CIRCUIT

Fig. 5-3 shows an RC circuit connected across a voltage source. The time required to charge the capacitor to 63.2% of the applied voltage is known as the time constant of the circuit. The value of this time constant in seconds equals the product of the circuit resistance in ohms and the capacity in farads. It may be found by using any of the following relations:

1. R (in ohms) × C (in farads) = t (in seconds).
2. R (in megohms) × C (in microfarads) = t (in seconds).
3. R (in ohms) × C (in microfarads) = t (in microseconds).
4. R (in megohms) × C (in picofarads) = t (in microseconds).

For example, a 0.1-microfarad capacitor in series with a 100K-ohm resistor will take one-hundredth (.01) of a second or 10,000 microseconds to reach 63.2% of the applied voltage.

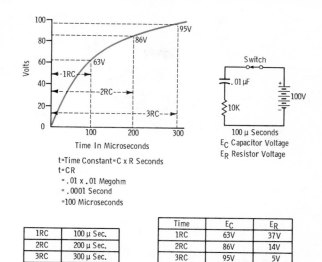

$$t = \text{Time Constant} = C \times R \text{ Seconds}$$
$$t = CR$$
$$= .01 \times .01 \text{ Megohm}$$
$$= .0001 \text{ Second}$$
$$= 100 \text{ Microseconds}$$

			Time	E_C	E_R
1RC	100 μ Sec.		1RC	63V	37V
2RC	200 μ Sec.		2RC	86V	14V
3RC	300 μ Sec.		3RC	95V	5V
4RC	400 μ Sec.		4RC	98V	2V
5RC	500 μ Sec.		5RC	99V	1V

Fig. 5-3. Time constants of an RC circuit.

In Fig. 5-3, a .01-μF capacitor is in series with a 10K-ohm (.01 megohm) resistor. Multiplying the capacitance (.01 μF) by the resistance (.01 megohm) gives a time constant of .0001 second, or 100 microseconds, for this circuit. This means that after 100 microseconds have elapsed, 63.2% of the applied voltage is across the capacitor, and 36.8% is across the resistor.

Since the applied voltage is 100 volts, the capacitor charge will be approximately 63 volts. Because of the charging current, the IR drop across the resistance will be approximately 37 volts.

In twice the time, or 200 microseconds, 63.2% of the remaining 37 volts is added to the original 63.2% charge. Approximately 86 volts will be across the capacitor, and approximately 14 volts will be across the resistor. Or:

$$200 \text{ microseconds} = 2RC$$
$$= 63 \text{ volts} + (63.2\% \times 37)$$
$$= 86.4 \text{ volts.}$$

This value can be found by following the E_C curve in Fig. 5-3.

Theoretically, the capacitor never reaches a fully charged condition. After 5 time constants, approximately 99% displacement of voltage across the circuit has occurred. For all practical purposes, this is sufficient to be considered a full charge. (Refer to the chart in Fig. 5-3.)

The time required to discharge a capacitor through a certain resistance is the same as the time required to charge it through the same re-

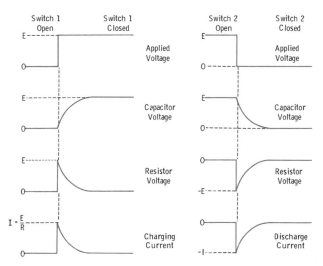

Fig. 5-4. RC charge and discharge curves.

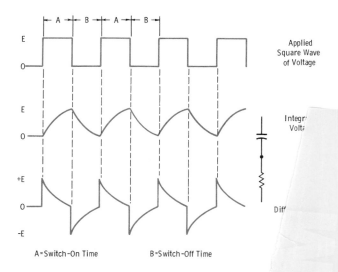

Fig. 5-6. Application of square wave of voltage to R

sistance. Therefore, the time constant is proportional to the time required to charge or discharge a capacitor.

In 1 time constant of the discharge period, 36.8% of the original charge will remain in the capacitor. The charge and discharge curves are shown in Fig. 5-4. Note the similarity; exponentially, they are the same.

FORMATION OF SQUARE AND SAWTOOTH WAVES

If a source of dc voltage connected to a resistive load is switched on and off in equal alternate periods, the applied electrical pressure across the resistor will be a symmetrical square wave of voltage (Fig. 5-5).

On the other hand, if the circuit is switched on and off in unequal alternate periods, the applied voltage to the load will be an asymmetrical square wave (Fig. 5-5). Therefore, by mechanically operating an on-off switch, we can generate two types of voltage waveforms. They are:

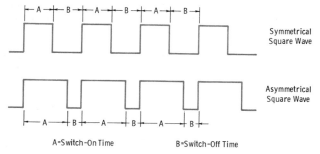

Fig. 5-5. Square waveforms.

1. Symmetrical square wave of voltage.
2. Asymmetrical square wave of voltage.

Now, if a fairly large capacitor is connected in series with the resistor, and a dc source of supply is switched on and off in equal time periods to produce an applied square wave of voltage, the resistive and capacitive components of the circuit will produce the following waveshapes (Fig. 5-6):

1. The capacitor voltage, known as the integrator voltage, will appear as a back-to-back sawtooth.
2. The voltage drop across the resistor, known as the differentiator voltage, will appear as a partially distorted square wave.

The polarity of the integrator voltage is unchanged during the charge and discharge period. The differentiator voltage is driven in two directions, positive and negative.

Increasing or decreasing the value of the capacitor in the RC network will change the integrator and differentiator voltage waveforms (Fig. 5-7). Note that the output waveforms for the 100-microsecond circuit are similar to the ones in Fig. 5-6. When the capacitor is increased to give a time constant of 1000 microseconds, only a slight voltage is obtained across the capacitor. The voltage across the resistor is distorted very little from the applied waveform. On the other hand, when the value of the capacitor is reduced to give a time constant of 10 microseconds, the voltage waveform across the capacitor is similar to the applied voltage. The waveform across the resistor

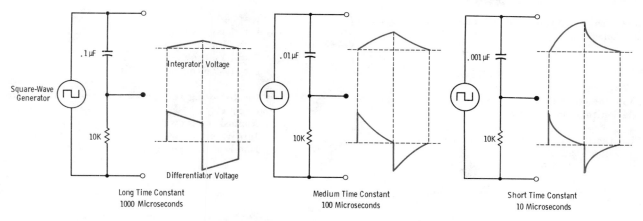

Fig. 5-7. Examples of time constants and their RC waveforms.

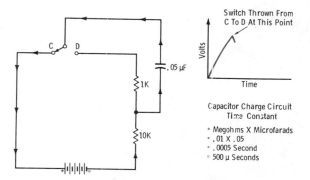

Fig. 5-8. Capacitor charge circuit for sawtooth waveform.

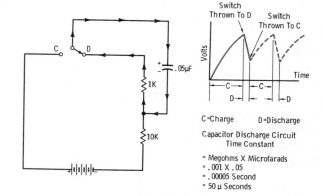

Fig. 5-9. Capacitor discharge circuit for sawtooth waveform.

is greatly differentiated and gives sharp positive and negative peaks. Because the waveforms applied to each of the circuits in Fig. 5-7 have the same frequency and amplitude, a square wave can be differentiated or integrated a variable amount to give the desired waveshape by properly selecting the values in the RC circuit.

At this point, we are interested in the integrator voltage. So, as we advance still further in the study of waveshapes and circuit analysis, let us refer to Fig. 5-8. Here we have a circuit wherein a capacitor is permitted to charge through a 10K-ohm resistor. By means of a switch, the capacitor is discharged through a 1K-ohm resistor. If the charge time is longer than the discharge time (for instance, 10 times longer), the charge and the discharge voltages of the capacitor will be a sawtooth of voltage (Fig. 5-9).

The slow charge and rapid discharge effects can be clearly seen. To obtain linear horizontal and vertical scanning for building a frame or raster, linear sawtooth waves must be generated and applied to the deflection systems of the cathode-ray tube.

QUESTIONS

1. What is the condition of a charged capacitor? What is the condition of a discharged capacitor?

2. When a potential is placed across an RC circuit, what happens to the voltage across the capacitor and resistor as time progresses?

3. Are the charge and discharge curves of a capacitor linear or nonlinear? What types of curves are these?

4. When the applied voltage is removed from an RC circuit and the fully charged capacitor is allowed to discharge through the resistor, what happens to the voltage across the capacitor and the resistor?

5. Define the time constant of an RC circuit. What basic equation is used to find the value of the time constant?

6. When a square wave is applied across an RC network, what is the capacitor voltage called? The resistor voltage?

7. As the charge time constant of an RC circuit is increased by changing the capacitance, what happens to the voltage

across each component when a square wave input voltage is applied?

EXERCISES

1. Show the charge and discharge curves of a capacitor in relation to voltage and time.

2. Find the time constant of an RC circuit when:
   (a) R = 100K ohms.
       C = .005 microfarad.
   (b) R = 1 megohm.
       C = .01 microfarad.
   (c) R = 1.5 megohms.
       C = 1000 picofarads.

3. Show the following voltage waveshapes in an RC circuit in time sequence:
   (a) Applied square wave of voltage.
   (b) Integrator voltage.
   (c) Differentiator voltage.

CHAPTER 6

Sawtooth Generators

Production of a sawtooth waveform usually involves the charge and discharge of a capacitor through resistors which differ greatly in value between the charge and discharge circuits. An introduction to this concept was discussed previously in Chapter 5. It has been shown that, to produce a sawtooth waveform, we need a simple circuit consisting of a source of voltage; a single-pole, double-throw switch; resistors; and a capacitor. The capacitor is charged through a high value of series resistance. The voltage across the capacitor at any instant has been shown in Fig. 5-1. Note that the initial portion of this voltage-versus-time curve is essentially a straight line. If we can short-circuit the capacitor before extreme curvature of the charge waveform has occurred and immediately initiate another charging cycle, we have produced a sawtooth wave. This sequence of events could be accomplished with a mechanically operated switch (Fig. 5-9). Since this entire operation must occur in a few millionths of a second, such a mechanically operated switch is obviously impractical. For this reason, we will resort to some electronic means to accomplish the switching sequence.

Although modern television receivers employ vacuum-tube or transistor oscillators and wave-shaping circuits to produce the ideal sawtooth scanning motion, it will be instructive to examine some earlier forms of circuits.

NEON-TUBE OSCILLATOR

The familiar neon-gas tube employed in sign lighting is one of the simplest automatic switches for short-circuiting a capacitor at the proper instant to produce a sawtooth voltage wave. A neon-filled tube, having a pair of electrodes and connected to a source of electrical potential, exhibits interesting properties as the voltage across the electrodes is gradually increased. No electrons will flow through such a tube until the voltage reaches a value known as the ionization potential. Until this voltage has been reached, the tube acts as an open circuit or as an extremely high resistance. However, when the ionization potential has been reached, the voltage removes electrons from the atoms of the gas and leaves these atoms (ions) with a positive potential. The free negative electrons are rapidly collected by the positive electrode. The positively charged ions are correspondingly attracted to the negative electrode, and current passes through the tube. The resistance of the tube suddenly changes, and it can be considered a voltage-operated switch. Fig. 6-1 shows how such a gas-discharge tube acts as a switch across a capacitor.

When the charge on the capacitor has produced a voltage across the tube equal to its ionization potential, the tube will suddenly conduct and start to discharge the capacitor. Once current conduction has started, it will continue even though the voltage has dropped below the original ignition point. Conduction will continue until a lower voltage level, known as the deionization potential, is reached. At this point, the tube returns to its nonconducting condition.

The charging cycle from the voltage source through the series resistor is resumed, and the cycle continues until ionization again occurs. This sequence of events is diagrammed in Fig. 6-1. Such an automatically operated switching circuit is more commonly known as a relaxation oscillator circuit.

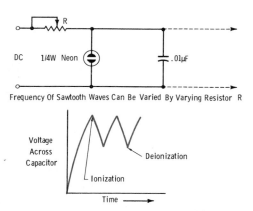

Frequency Of Sawtooth Waves Can Be Varied By Varying Resistor R

Fig. 6-1. Neon-tube relaxation oscillator.

THYRATRON OSCILLATOR

The grid-controlled thyratron is an improved form of gas-discharge tube. Such a tube acts essentially the same as the simple neon lamp previously described, with the following exceptions:

1. A source of electrons from an electrically heated cathode supplies the electrical current for the discharge portion of the cycle.
2. The triggering action is controlled by an additional element similar to the grid of the familiar vacuum tube. This element is normally held at a negative potential and prevents current conduction between the cathode and plate by its repelling action on the electrons emitted by the cathode.
3. The gas normally employed is mercury or argon instead of neon.

A relaxation oscillator can be built with this tube; it is easier to control than a simple neon gas tube. Fig. 6-2 shows the basic circuit of a

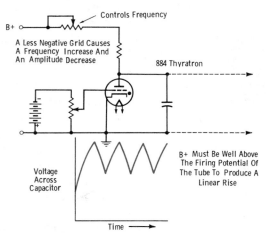

Fig. 6-2. Thyratron relaxation oscillator.

thyratron sawtooth generator. To ensure that the rise of the charging potential is linear, only a small portion of the B+ voltage is allowed to charge the capacitor. The grid is held at a sufficiently negative potential to ensure that there is no plate current. In the diagram in Fig. 6-2, this grid potential is provided by a bias battery. Tripping of the circuit can be produced by a positive pulse applied to the grid. Once initiated, plate current will continue until the plate voltage has dropped to a point which corresponds to the deionization potential for the neon-gas tube.

The thyratron oscillator circuit (Fig. 6-2) was employed in some cathode-ray oscilloscopes and was also used in some early television sets, both in this country and abroad. Television receivers no longer employ gas-tube relaxation oscillators, mainly because such oscillators are not sufficiently reliable in operation with fluctuating temperature and power-line voltage.

VACUUM-TUBE SAWTOOTH GENERATORS

We have seen how a voltage is applied to a capacitor through a series resistance to produce a sawtooth voltage waveform. After the capacitor has reached a predetermined charge, the voltage is removed by a switching action. This action can be more readily accomplished with thyratrons rather than with neon tubes.

Television sets have generally employed three types of circuit arrangements, or combinations of these circuits, to produce sawtooth waveforms.

1. *The multivibrator.* This circuit arrangement has many variations. The most popular variation is the cathode-coupled version.
2. *The blocking oscillator.* This type of circuit permits the formation of a short pulse of energy. This pulse can be used to produce the sawtooth wave across a capacitor directly associated with the oscillator tube, or the pulse can trigger a discharge tube which acts as a switch across the capacitor.
3. *The sine-wave oscillator.* An oscillator of the correct frequency supplies the timing voltage for the discharge tube. The sine-wave output of this oscillator is modified into short pulses by waveshaping circuits. These pulses then operate a discharge tube to produce sawtooth waves.

Multivibrators

The most popular sawtooth generator in tube-type receivers is the multivibrator. The multivibrator is another form of relaxation oscillator and employs vacuum tubes, resistors, and capacitors in a feedback arrangement.

The multivibrator is useful because tubes can act as automatic switches to control the charge and discharge of capacitors. This action produces a sustained output of rectangular waveform, the frequency of which can easily be controlled by the horizontal or vertical synchronizing pulses.

Several versions of the multivibrator circuit are found in modern television sets. Since they are derived from a basic or conventional type, we will first examine the theory and operation of the fundamental circuit.

The Conventional Multivibrator — The basic free-running multivibrator can be considered a two-stage resistance-capacitance coupled amplifier. Overall feedback is applied by means of a capacitor, connected from the output of the second stage to the input of the first stage.

Fig. 6-3A shows a familiar two-stage audio amplifier. The added capacitor C1 (shown in dotted lines) converts this amplifier into a free-running

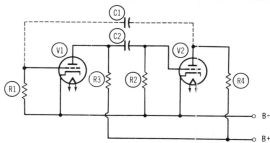

(A) Two-stage amplifier with capacitive feedback.

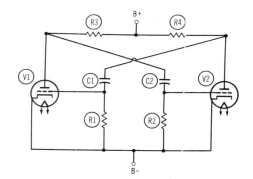

(B) Same circuit as (A), showing symmetry of circuit.

Fig. 6-3. The conventional multivibrator.

multivibrator. Fig. 6-3B shows a symmetrical rearrangement of the same circuit.

In a single-stage resistance-coupled amplifier, the plate voltage is 180° out of phase with the input voltage. Therefore, the output voltage of the second stage (V2) of the two-stage amplifier will again have been inverted and will be in phase with the input voltage to V1. Capacitor C1 of Fig. 6-3 will impress a voltage of the proper polarity upon the first-stage grid to increase the original input voltage, and oscillation can take place.

To show how oscillation starts and is maintained in this circuit, let us assume that the cathodes are heated and that B+ voltage is applied. Both grid circuits are returned to their respective cathodes through grid resistors. At the instant B+ potential is applied, the grids will be at cathode potential, or zero bias. Grid- and plate-current conduction will start in each tube.

It will be instructive at this point to list the sequence of events which produces the sustained rectangular-shaped output wave of the multivibrator. (Refer to Figs. 6-3 through 6-6.)

1. Since the resistance of the internal cathode-to-grid path under this initial zero bias and high grid current is much lower than the resistance of grid resistors R1 and R2, capacitors C1 and C2 will begin charging from the B+ supply through resistors R4 and R3, respectively. This charging path is shown by the arrows in Figs. 6-4A and 6-4C.

2. If the characteristics of both tubes and the value of the circuit elements were exactly matched, the charging rate of both capacitors would be identical, and the plate currents of both tubes would rise simultaneously. A state of equilibrium would be reached, and the circuit would not oscillate. Since such conditions are not met in practice, a balance is not established.

3. Actually, one of the tubes will start to conduct plate current sooner than the other. This conduction could be caused by lower plate resistance, hotter cathode, or a slightly lower value plate-load resistor. Let us assume that the plate current of V1 has started to rise a fraction of a second ahead of the plate current of V2.

4. This rise of plate current will be accompanied by a drop in plate-to-cathode resistance and by a corresponding drop in plate-to-

Arrows Show Direction
Of Electron Flow During
Charge And Discharge

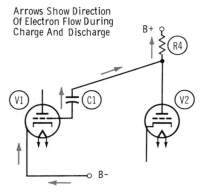

(A) Charge path of C1.

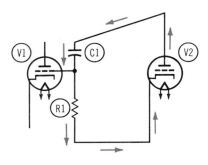

(B) Discharge path of C1.

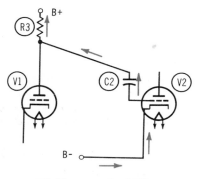

(C) Charge path of C2.

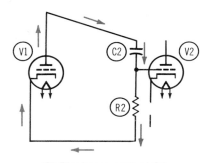

(D) Discharge path of C2.

Fig. 6-4. Capacitor charge and discharge paths in a multivibrator.

cathode voltage. Fig. 6-5A shows this set of operating conditions. The low plate resistance of V1 forms a discharge path for C2, as shown in Fig. 6-4D.

5. The discharge current of C2, flowing through the high value of grid resistor R2, develops a high negative grid bias on the grid of V2. This bias drives the tube beyond plate-current cutoff, as shown in Fig. 6-5B. The bias developed by this discharge can be as high as 30 to 50 volts in the example shown.

6. Since the plate current of V2 has been cut off, its plate-to-cathode voltage becomes that of the B+ supply (Fig. 6-5B). The plate-to-cathode voltage will remain at that value until the grid voltage has reached the point where the grid is no longer cut off.

NOTE: Since the foregoing conditions have brought the cycle of operation to one of the two stable, or relaxed, operating points of the circuit, it would be helpful to summarize the changes of circuit voltages which have occurred

over the period covered by Steps 1 through 6.

V1—Plate-to-cathode voltage at its minimum value and steady. Tube conducting. Control-grid voltage zero and steady.

V2—Plate-to-cathode voltage at its maximum value and steady. Tube not conducting. Control-grid voltage highly negative, but falling exponentially with time as C2 discharges through R2.

7. The time required for C2 to discharge will depend on the time constant of the discharge circuit of C2, R2, and the plate resistance of V1. (See Fig. 6-4D.) The negative voltage across R2, which constitutes the grid bias of V2, finally becomes low enough to allow V2 to conduct heavily. Fig. 6-6 shows the waveforms of the grid and plate voltages of both tubes as a function of time. The parts of the waveforms between (A) and (B) in Fig. 6-6 cover the steps outlined up to this point.

8. As V2 starts to conduct, conditions in this tube become identical to those in Step 4 for

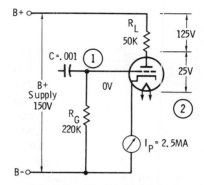

① Grid At Zero Bias After C Regains Charge By Grid Conduction Through Tube.

② r_p Low (10,000 Ω Approx.) Plate Current Maximum.

(A) Tube conducting.

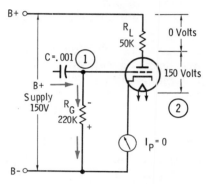

① Grid Biased Negative By Discharge Of C Through R_G. I_p Cut off.

② r_p Of Tube Very High; Practically Infinite. Plate Current Zero.

(B) Tube not conducting.

Fig. 6-5. Tube operating conditions in a multivibrator.

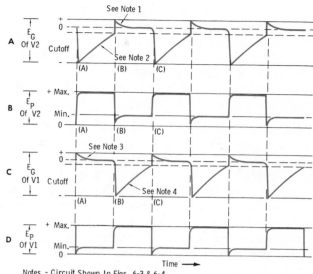

Notes - Circuit Shown In Figs. 6-3 & 6-4
1. C2 Charge Path - See 6-4C
2. C2 Discharge Path - See Fig. 6-4D
3. C1 Charge Path - See Fig. 6-4A
4. C1 Discharge Path - See Fig. 6-4B

Fig. 6-6. Typical waveforms of a symmetrical multivibrator, showing square-wave switch action.

V1, except that the tubes and capacitors have exchanged functions, and the discharge path of capacitor C1 is now as shown in Fig. 6-4B.

9. The discharge current of C1 flowing through R1 now biases V1 beyond cutoff, as described in Step 5.
10. Since the plate current of V1 has been cut off, the plate-to-cathode voltage assumes the value of the B+ supply, similar to V2 in Step 6.
11. The rise in plate voltage of V1 is impressed on capacitor C2, starting the charging cycle shown in Fig. 6-4C.
12. Since the internal grid-to-cathode path of V2 is conductive because of the zero grid bias, the charging resistance is small, and

C2 is rapidly charged. This action is shown at time (B) in Fig. 6-6.

We can now summarize the conditions of circuit voltage and compare them with those found at the end of Step 6.

V1—Plate-to-cathode voltage at its maximum value and steady. Tube not conducting. Control-grid voltage highly negative, but falling exponentially with time as C1 discharges through R1.

V2—Plate-to-cathode voltage at its minimum value and steady. Tube conducting. Control-grid voltage zero and steady.

Note that the new conditions, which represent the other stable, or relaxed, operating point are the same as before, except that the tubes and grid circuits have changed places. This cycle of events is shown in the waveform diagrams of Fig. 6-6 between times (B) and (C). This square-wave generation will continue at a frequency determined by the charge and discharge time constants of coupling networks R1-C1 and R2-C2.

In this symmetrical circuit, it has been assumed, but not stated, that the corresponding grid resistors, plate resistors, and coupling capacitors are equal. When this is true, the time constants are equal, and the output waveforms from the plates are identical. The frequency of this multivibrator can be changed by altering either resistors R1 and R2 or capacitors C1 and C2. A lower time constant will increase the frequency. If the values of R or C are changed equally, the output wave remains symmetrical.

The Asymmetrical, or Unbalanced, Multivibrator—To produce the type of sawtooth wave required for television scanning with a multivibrator, succeeding square waves must be unequal in length or spacing. For this reason, the time constant of the RC circuit of one tube is made much greater than the time constant of the other. Such a multivibrator is called asymmetrical. Fig. 6-7 shows the waveforms obtained when the circuit constants of the symmetrical multivibrator just described are changed so that the product (R1 × C1) in the grid circuit of V1 is much smaller than the product (R2 × C2) in the grid circuit of V2.

Waveform D of Fig. 6-7 shows a short pulse of plate current occurring in V2 once each cycle. We will employ this pulse to produce the scanning sawtooth in proper time relationship to the scanning of the camera tube at the transmitter.

Use of the Multivibrator to Produce Sawtooth Scanning—Fig. 6-8 shows a circuit similar to those discussed for symmetrical and asymmetrical multivibrators. By adding two new circuit elements, we can generate sawtooth voltage waves to control the electron beam with either the horizontal or the vertical sweep circuits. These new circuit elements are C3 and C4 (Fig. 6-8). Coupling

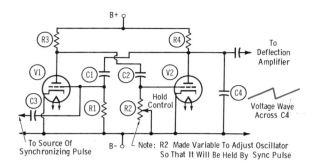

Fig. 6-8. Asymmetrical multivibrator, showing input-coupling and sawtooth-forming capacitors.

capacitor C3 connects the multivibrator circuit to a source of synchronizing pulses, which are part of the transmitted television signal. The function of the synchronizing pulses and how they control the frequency of the multivibrator will be covered in a later chapter.

At this time, the additional circuit element which concerns us is capacitor C4 between the plate and cathode of V2. For a horizontal-line scanning frequency of 15,750 Hz, the circuit is so arranged that the time constant of R1 and C1 is approximately one-ninth the time constant of R2 and C2. The plate current of V2 will consist of short pulses (Fig. 6-7) which represent low resistance. During the time shown as the conducting period, V2 will act as a short circuit across capacitor C4. The multivibrator thus acts as a periodic switch and fulfills the requirements covered previously for producing a sawtooth wave.

A significant difference between the circuit in Fig. 6-8 and the one in Fig. 6-3 is that R2 has been made variable. This variable resistor is one of the major controls of a television receiver. From the previous discussion of multivibrator theory, we know that varying R2 alters the length of the portion of the operating cycle controlled by R2 and C2 in Fig. 6-7. This represents the active portion of the sawtooth wave when the face of the cathode-ray tube is scanned during video modulation. This variable adjustment permits the multivibrator to be locked in with the synchronizing pulse and is known as a hold control.

The voltage of the sawtooth wave across capacitor C4 (Fig. 6-8) is too small to produce the required deflection of the electron beam; therefore, amplifiers are needed.

Cathode-Coupled Multivibrator—A variation of the multivibrator is the cathode-coupled circuit. This circuit is shown in Fig. 6-9. A significant dif-

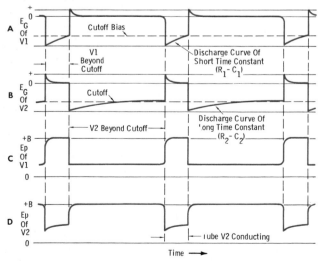

Fig. 6-7. Waveform of an asymmetrical, or unbalanced, multivibrator.

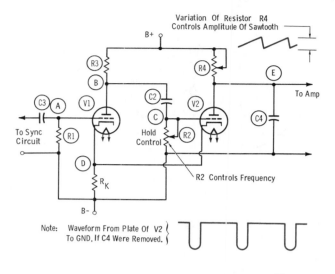

Variation Of Resistor R4
Controls Amplitude Of Sawtooth

B+

R3
B
C3 A V1
To Sync
Circuit
R1
D
R_K
B-

R4
C2
C
Hold
Control
R2
E
To Amp
C4

R2 Controls Frequency

Note: Waveform From Plate Of V2
To GND, If C4 Were Removed.

See Fig. 6-10 For Voltage Waveforms At Points Indicated Thus: (A)

Fig. 6-9. A cathode-coupled multivibrator.

ference between this circuit and the conventional multivibrator is that feedback is accomplished in two ways. Coupling capacitor C2 transfers charges from the plate of V1 to the grid of V2. In addition, this circuit employs a cathode-bias resistor common to V1 and V2. This common-cathode resistor is responsible for the unique action af the circuit. The second tube (V2) functions as a switch or discharge tube for capacitor C4, which produces the sawtooth waveform. From the theory of the conventional and asymmetrical multivibrators, as previously discussed, the action of this cathode-coupled version can be readily understood. We will again assume that the cathodes of the tubes are heated and that B+ potential is applied. Let us follow the sequence which allows this circuit to generate asymmetrical pulses.

1. Capacitor C2 will charge through R3 and the grid to the ground (or B−) circuit of V2. This action occurs quite rapidly, since the grid of V2 is initially at zero potential.
2. Plate-current conduction will start in both V1 and V2. Bias voltage for both of these tubes will be developed across cathode resistor R_K.
3. This bias voltage will immediately start to decrease the plate current of both tubes, which were initially in a conductive condition since the control grids were at zero potential.

4. The conduction of plate current through V1 causes a lower plate-to-cathode voltage drop and a corresponding lower plate resistance in this tube.
5. The low-resistance path of V1 initiates the discharge of coupling capacitor C2 through R2 and R_K. As in the conventional multivibrator, the current through R2 produces a high negative bias on the control grid of V2, which immediately drives the tube beyond its plate-current cutoff point. Note that this circuit differs from the conventional multivibrator; there is no coupling capacitor between the plate of V2 and the grid of V1. For this reason, the plate current of V2 is immediately cut off.

As in our discussion of the conventional multivibrator, it will be constructive to summarize the voltage conditions which have occurred up to this point:

V1—Plate-to-cathode voltage at its minimum value. Tube conducting. Control-grid voltage negative and steady. The tube is self-biased by its own plate current through common cathode resistor R_K.

V2—Plate-to-cathode voltage rising along the linear portion of the charging curve. This rise of plate voltage charges capacitor C4 and initiates the first part of what will eventually become a sawtooth wave of voltage. Control-grid potential highly negative and exponentially diminishing in value.

6. V2 has been cut off during this period. Capacitor C2 has been discharging through R2, R_K, and the cathode-to-plate circuit of V1. It is interesting to note that, as the plate current of V2 was cut off by the high negative bias produced across R2 by the discharge of C2, the plate-to-cathode voltage did not immediately assume the value of the B+ supply. This is because the plate voltage of V2 was maintained by the charge of C4, which started simultaneously with the application of B+ voltage. For this reason, the plate voltage waveform of V2 will be sawtooth, instead of rectangular, because of the charge flowing into C4. If C4 is removed from the circuit, the plate-to-cathode voltage of V2 would rise immediately to the B+ value, since the grid of this tube is cut off by the high negative voltage resulting from the discharge of C2.

7. As in the other types of multivibrators already discussed, when the bias on V2 falls to a value equal to the grid cutoff potential, V2 will start to conduct.

8. When conduction occurs in V2, C4 will be rapidly discharged through the plate-to-cathode circuit of this tube, and its voltage will drop to a minimum value (Fig. 6-10E). The sequence of events at this point has resulted in a sawtooth wave of voltage across capacitor C4. Thus far, the circuit is like the conventional and asymmetrical multivibrators previously discussed.

9. A different action now takes place. The sudden pulse of plate current, which occurs when V2 conducts, flows through cathode resistor R_K. Since this resistor is common to the cathode circuits of both V1 and V2, the voltage produced by this plate-current pulse immediately drives the grid of V1 negative with respect to its cathode.

10. This negative bias suddenly increases the plate-to-cathode resistance and the plate-to-cathode drop of V1. The sudden increase in plate voltage charges C2. A positive voltage is instantaneously impressed on the grid of V2; and the plate current pulse of V2, which started the cycle, is momentarily increased.

11. The cumulative increase of plate current through common-cathode resistor R_K finally biases the grid of V1 enough to completely cut off the plate current, and the plate-to-cathode voltage of V1 rises to its maximum value.

12. Capacitor C2 has become charged, and the plate current of V2 relaxes.

13. This decreased current in R_K reduces the bias of V1. As V1 conducts, C2 is discharged through R2, R_K, and the plate circuit of V1. The current through R2 drives the grid of V2 to cutoff, and the cycle is repeated.

Because a sudden and cumulative action was produced in the circuit by the coupling of tubes V1 and V2 through a common-cathode resistor, this circuit is called a cathode-coupled multivibrator. The cathode-coupled multivibrator is preferred in television over the asymmetrical multivibrator because:

1. It can be triggered and controlled by a negative pulse of voltage; thus, the control circuits can often be simplified.

2. Its sudden and cumulative pulsing action in V2 permits a higher ratio of linear-sweep time to return time.

3. Variable resistors R2 and R4 of V2 (Fig. 6-9) permit control of its scanning frequency and amplitude.

The waveshapes of the voltages at various points in this circuit as a function of time are shown in Fig. 6-10. These waveshapes are identified in Fig. 6-9 by letters enclosed in circles. Voltages are always measured to B− or ground.

Blocking Oscillators

Another type of vacuum-tube circuit for producing controlled sawtooth voltages is the blocking oscillator. The blocking oscillator was originally quite popular in tube-type receivers, especially as a vertical-sweep oscillator, but has been discontinued in favor of the multivibrator.

Fig. 6-11 shows a simple blocking oscillator. Upon casual inspection, it looks like a Hartley oscillator with an iron-core transformer. Basically, it is such an oscillator. However, instead of sustained sine-wave oscillations, it produces short pulses of energy, with correspondingly long in-

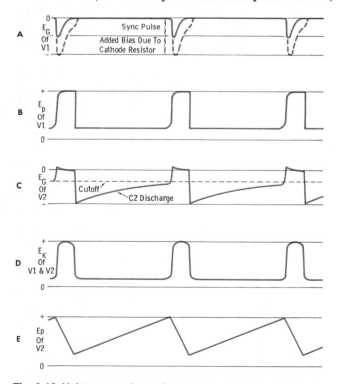

Fig. 6-10. Voltage waveforms in cathode-coupled multivibrator.

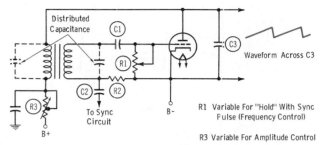

Fig. 6-11. Blocking oscillator.

R1 Variable For "Hold" With Sync Pulse (Frequency Control)

R3 Variable For Amplitude Control

tervals of relaxed action. For this reason, it is classified as another form of relaxation oscillator. Two significant differences distinguish this circuit from the common Hartley oscillator:

1. The time constant of grid resistor R1 and grid capacitor C1 is such that long periods of blocked plate current occur between short periods of plate-current conduction. During these short conductive periods, oscillation takes place.

2. The natural period of oscillation of the transformer, with its associated distributed and lumped circuit capacitances, is such that the desired pulse time approximates one-half cycle of the frequency at which the circuit would oscillate if it were the continuous sinusoidal type.

As was done with the other types of sawtooth oscillators, it will be instructive to follow through, in sequence, the various actions which take place in this circuit.

1. We will again assume that the cathode of the tube is heated and that the plate circuit is suddenly closed to provide B+ potential. Since the grid is initially at cathode or zero potential, the tube will start to conduct plate current which will pass through the transformer primary. This sudden rush of current will set up a magnetic field in the core of the transformer, and a secondary voltage will be induced across the grid winding. The direction of these windings is such that the primary current will cause a positive potential to appear at the grid with respect to the cathode or ground.

2. The positive voltage applied to coupling capacitor C1 makes the grid more positive than the cathode. The grid then attracts electrons from the emitted cathode current, and grid current flows through resistor R1.

3. Simultaneously, the increasing positive grid potential makes the plate draw still more current, until plate-current saturation is reached. When the plate current reaches a steady maximum value, no further change of current occurs in the primary winding of the transformer.

4. Since the voltage induced into the secondary depends on the change of magnetic flux, the secondary voltage of the transformer will cease to rise.

5. As the grid becomes less positive (C1 discharging through R1), the plate current through the primary falls, and the magnetic field linking the secondary coil collapses. The time taken for this sudden rise and fall of grid voltage is governed by the natural resonant frequency of the transformer and its associated circuit capacitances.

6. The collapsing field in the transformer due to the dropping plate current induces a secondary voltage. This secondary voltage is opposite to the voltage produced by the original plate-current pulse. Capacitor C1 discharges through resistor R1. The grid is driven more and more negative. The plate current quickly falls until it finally reaches a cutoff point. Although the reversal of grid voltage and the cutoff of plate current has taken considerable time to describe, the action is practically instantaneous.

7. From this point, the action in the tube follows the action in the multivibrator. The grid potential follows an exponential curve of RC discharge until plate conduction is again reached. The waveform of grid voltage is shown in Fig. 6-12A.

8. The time taken for the discharge of capacitor C1 depends on the time constant (R1 + R2) × C1.

9. As the tube starts to conduct again, oscillation begins and the cycle is repeated. From the curve of Fig. 6-12B, we see that the plate voltage of the tube is nearly steady and is at the B+ value between these oscillatory pulses. We have fulfilled the conditions of sawtooth charge and discharge of capacitor C3 in Fig. 6-11 and have, therefore, produced a sawtooth scanning voltage. As in previous circuits, grid resistor R1 can be made variable for use as a frequency control (hold control), and a plate-circuit resistor

can be made variable for use as a width or height control.

In the circuit in Fig. 6-11, note that the saw-tooth-generating capacitor C3, connected from plate to cathode of the tube, has modified the shape of the plate-voltage waveform. A similar action occurred in the cathode-coupled multivibrator circuit in Fig. 6-9. Without capacitor C3, the voltage waveform between the plate and cathode of the tube would look like the wave in Fig. 6-12B. The amplitude of the wave above the B+ axis at point X is caused by the energy stored in the transformer primary.

When capacitor C3 is connected, the voltage wave from plate to cathode assumes the shape shown in Fig. 6-12C. This is the desired sawtooth deflection waveform except for the distorted section at point Y. The sudden rise of the curve at this point is due to the additional charging voltage when the magnetic flux of the transformer collapsed, as explained in the foregoing portion of the text. The plate voltage could not follow the curve in Fig. 6-12B because of the terminal voltage of capacitor C3. The distorted section of the wave at point Y is actually of no consequence because it is blanked out, as will be explained later.

Sine-Wave Generators

Sine-wave generators are biased to run in a fashion similar to Class-C transmitter operation, where plate current is cut off for part of the cycle. Only a short portion of the sine wave is used. It is passed through clipping stages to "bite off" a small section of the wave. The output of the clipper is a pulse. Part of the necessary clipping of the sine wave to produce a pulse is already accomplished in the oscillator itself. The usefulness of this type of circuit will be discussed later when we consider how scanning is controlled by synchronizing pulses.

TRANSISTOR SAWTOOTH GENERATORS

The sawtooth-forming circuits that have been discussed up to now have employed vacuum tubes as automatic switches to control the charge and discharge of capacitors. The transistor performs very well as a switch and has replaced the vacuum tube in sawtooth-forming circuits.

The most popular transistor-type sawtooth generators presently used in television receivers are

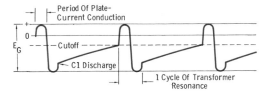

(A) Grid-voltage waveform.

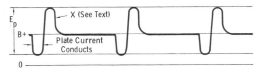

(B) Plate-voltage waveform which would occur if C3 (Fig. 6-11) is disconnected.

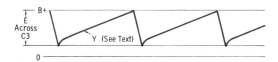

(C) Sawtooth waveform across C3.

Fig. 6-12. Voltage waveforms in a blocking oscillator.

the multivibrator and the blocking oscillator. The multivibrator is usually found in the vertical sweep section; the blocking oscillator is commonly employed in the horizontal sweep section. A sine-wave oscillator is used in some transistor horizontal sweep circuits.

Multivibrators

The operation of a transistor-type multivibrator circuit is very similar to the tube-type multivibrator discussed earlier in this chapter. A simplified version of a multivibrator circuit employing transistors is shown in Fig. 6-13. This is basically a two-stage transistor amplifier with the output of the second stage coupled back to the base of the first stage, by way of capacitor C2.

When the B+ voltage is first applied to the circuit, capacitor C1 will begin to charge through the base-emitter junction of Q2 and resistor R2. Similarly, capacitor C2 will charge through the

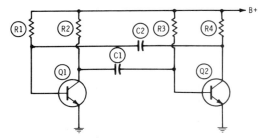

Fig. 6-13. Basic multivibrator circuit using transistors.

base-emitter junction of Q1 and R4. The charging current through the base-emitter junctions of the transistors will cause them to begin conduction. As in the case of the tube-type multivibrator, one of the transistors will start to conduct sooner than the other. For this example, let us assume that Q2 begins to conduct before Q1.

As Q2 starts to conduct, there will be a voltage drop across R4, which will cause the collector voltage of Q2 to decrease. The rapid decrease in collector voltage of Q2 will cause capacitor C2 to discharge through R1 and R4. This action will quickly cut off transistor Q1 and cause its collector voltage to rise to the value of the B+ voltage. The voltage rise at the collector of Q1 will cause the charging action of capacitor C1 to increase, and transistor Q2 will conduct heavily.

Transistor Q2 will continue to conduct until C2 is fully discharged. The voltage polarity across C2 will become reversed when the capacitor is fully discharged and transistor Q1 will again be forward biased. As Q1 starts to conduct, its collector voltage will rapidly decrease due to the voltage drop across R2, and capacitor C1 will discharge through R2 and R3. Transistor Q2 will be

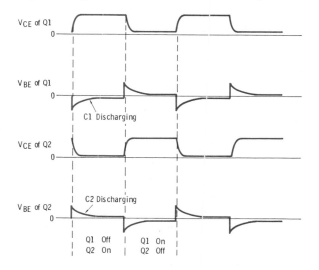

Fig. 6-14. Waveforms for transistor multivibrator circuit.

cut off, and capacitor C2 will begin charging through R4 and the base-emitter junction of Q1, which is now forward biased. Transistor Q1 will continue to conduct until capacitor C1 is completely discharged. Transistor Q2 will come out of cutoff, and the entire action will be repeated. The voltage waveforms for the multivibrator just described are shown in Fig. 6-14.

In actual practice, the values chosen for the resistors and capacitors in the multivibrator circuit are such that the output is asymmetrical. Again, this is necessary to produce the sawtooth waveform required for the scanning circuits in television receivers.

Blocking Oscillators

Although the blocking oscillator has been replaced for the most part by the multivibrator in vacuum-tube receivers, it is being used quite extensively in the solid-state receivers. Fig. 6-15 shows a typical blocking oscillator using a transistor as the active element. This particular circuit is used in the vertical sweep section.

In this circuit, the tightly coupled windings of transformer T1 provide feedback from the collector of transistor Q1 to the base. The transistor operates in a way similar to a Class-C amplifier in that it conducts only for brief intervals. During this conduction, the circuit produces pulse waveforms. The base of Q1 is driven strongly during conduction by the amplified output of the collector. The resultant negative-going pulses are rectified by the base-emitter junction and capacitor C5 is charged to a comparatively high negative dc potential. With the emitter more negative than the base, Q1 remains in a cutoff state until capacitor C5 discharges sufficiently through R6. As the transistor comes out of cutoff, another surge of oscillation occurs, causing Q1 to be blocked again. The setting of hold control R4 determines how long the transistor will be cut off, or blocked, since it controls the base bias.

Negative-going sync pulses are coupled to the base of Q1 through T1. Then, sync pulses trigger the blocking oscillator into conduction sooner than

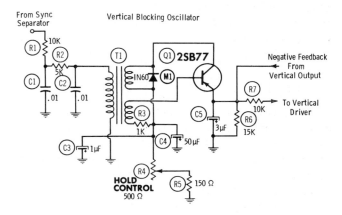

Fig. 6-15. Typical transistor blocking-oscillator circuit.

the circuit would otherwise come out of cutoff. In this way, synchronization of the blocking oscillator is accomplished. Diode M1 connected across one winding of T1 serves as a limiter to prevent the transformer from applying an excessive peak voltage to the collector of Q1. M1 also acts as a blocking diode to prevent coupling of the blocking oscillator waveform back into the preceding circuit.

The pulse output of transistor Q1 is modified into a sawtooth waveform by the same components that provide the time constant of the oscillator (C5 and R6). The discharge of C5 through R6 produces a basic exponential waveform, which is then linearized into the desired sawtooth waveform by negative feedback from the sweep output circuit.

All of the television receivers now being manufactured employ solid-state devices rather than vacuum tubes. As a result, a wide variety of oscillator circuits is used in these sets. In some instances, the active devices and other components for an oscillator circuit are contained within an integrated circuit (IC). Such an oscillator circuit may be a conventional multivibrator, as described previously, or it may consist of a single stage with feedback to sustain oscillation provided by the driver or output circuit. The driver stage may be included within the same IC as the oscillator.

A SUMMARY OF MULTIVIBRATORS AND BLOCKING OSCILLATORS

1. Multivibrator circuits employ vacuum tubes or transistors as electronic switches to control the charge and discharge cycles of a capacitor for the production of sawtooth waves.

2. The switching action by the tubes is practically instantaneous and is limited in timing by the circuit elements rather than by the tube itself. The velocity of electrons in a tube suddenly driven to saturation after plate-current cutoff can approach one-tenth the speed of light, or 18,600 miles per second.

3. When an electron tube acts as a voltage-operated switch, it suddenly changes from no plate current to where plate current is limited only by the ability of the cathode to supply electrons (plate-current saturation). These two conditions are illustrated

in Fig. 6-5. Since this action is also found in other sections of a television receiver, you should become thoroughly familiar with the steps involved.

4. Multivibrator circuits may be either symmetrical or asymmetrical. Symmetrical circuits produce square voltage waves. The symmetrical multivibrator is used in the television transmitter to produce the complex television signal. It is not used in the receiver; for that reason, we are interested only in the asymmetrical type.

5. The asymmetrical multivibrator produces voltage output waves in which a short rectangular pulse is followed by a long "gap." This short pulse of voltage can be used to switch a vacuum tube from plate-current cutoff to plate-current saturation, and back again to cutoff.

6. If a capacitor of the correct value is connected across the plate of the output stage, the long periods between pulses will be occupied by the gradual charging cycle, and the voltage across the capacitor will rise linearly.

7. When the short rectangular pulse of voltage is suddenly applied to the grid of the tube, the tube becomes conductive and short-circuits the capacitor, and a new sawtooth wave is started.

8. An important reason for using this means of generating sawtooth waves is that dual-triode tubes can be employed, instead of the two tubes shown in the illustrations.

9. Checking the values of resistors and capacitors, making sure that there are no open or short circuits, and substituting tubes constitutes the normal service procedure in tube-type multivibrator scanning generators.

10. The blocking oscillator requires only one tube or transistor, contrasted with the multivibrator which depends on a phase rotation of 360° through two tubes or transistors. The phase of feedback in the blocking oscillator is provided by the relationship of transformer windings.

11. The characteristics of the transformer determine the length of the conduction time or "closed switch" part of the cycle. This time is approximately one-half cycle of the resonant frequency of the transformer and its associated capacitances.

12. The "relaxed" time between "switch on" or conduction periods is determined by an RC time constant. This time constant can be made variable to control the frequency of oscillation and allow synchronization with the transmitted signal.

13. In the vacuum-tube blocking oscillator, plate-to-grid feedback through the transformer sets up a grid current. This grid current charges the grid capacitor, and the grid is momentarily driven positive. The plate current rises to its saturation value and then falls. As the induced grid voltage reverses, the grid capacitor discharges through the grid resistor. The grid is driven very negative, and plate current is cut off. The cycle repeats as soon as the voltage across the grid resistor reaches a value which will allow the grid to again initiate plate current.

QUESTIONS

1. Name two gas-filled tubes that can be used in oscillator circuits.

2. How is feedback accomplished in a conventional multivibrator circuit?

3. What controls the bias of a conventional multivibrator circuit?

4. What are the distinguishing characteristics of an asymmetrical, or unbalanced, multivibrator?

5. By what two ways does the circuit of a cathode-coupled multivibrator differ from that of a conventional tube-type multivibrator?

6. When the second stage of the cathode-coupled multivibrator conducts, what action does the sawtooth-forming capacitor across the output perform?

7. In a blocking oscillator, what happens in the transformer when plate current is increasing?

8. What happens in the transformer when the plate current decreases?

9. What type of waveform is produced by each of the following circuits in the absence of a sawtooth-forming capacitor:
    (a) Symmetrical multivibrator?
    (b) Asymmetrical multivibrator?
    (c) Blocking oscillator?

10. What two types of sawtooth generators are normally found in modern transistor television receivers?

EXERCISES

1. Draw the circuit of a conventional multivibrator.

2. Break down the circuit in Exercise 1 and show the charge and discharge paths of the coupling capacitors.

3. Draw the circuit of a cathode-coupled multivibrator. Include the input capacitor and the sawtooth-forming capacitor.

4. Draw the basic circuit of a transistor blocking oscillator.

CHAPTER 7

Sawtooth Generator Control and Production of Scanning Waveforms

At this time, let us examine in greater detail the sawtooth scanning waveforms which produce the raster. We have mentioned before that the horizontal and vertical sawtooth motion of the electron beam must keep in synchronization with similar sawtooth scanning movements occurring at practically the same instant in the camera tube at the transmitter.

To accomplish this synchronization, pulses which control the horizontal and vertical scanning are transmitted in the television signal. These pulses occur between each horizontal line. During the scanning of the line, the receiver is "on its own." However, during the short interval between successive horizontal lines, the deflection circuits of the receiver are the absolute "slave" of the transmitter if the set is well designed, operating properly, and being used in an area of adequate signal strength.

How the synchronization pulses are separated from the complex signal will be taken up later in this course. At this point, we will examine the relationship between the timing of these pulses and the control of the sawtooth scanning of the receiver. Fig. 7-1 shows the sequence of events during the scanning of one horizontal line and during the return of the electron beam to start the scanning of the next line.

The sawtooth line in Fig. 7-1 shows the desired linear trace and retrace motions of the electron beam in the tv picture tube. It does not necessarily depict the exact wave of current which must be passed through the deflection coils of the picture tube. As a matter of fact, we will find later that the deflection current must be distorted somewhat

to accomplish the linear sweep and rapid flyback of the beam of electrons tracing the picture.

Fig. 7-1 shows, therefore, the *ideal* sawtooth for controlling the horizontal scanning motion in a television receiver. At point A, the electron beam starts to cross the face of the picture tube horizontally from left to right. We will assume the picture tube is a 17-inch type and has an active picture width of 14 inches. The beam has been blanked out from A to B, and the picture starts at point B. Between points B and C, as the uniform motion progresses, the video modulation produces the picture.

As we have previously stated, the picture frame consists of 525 horizontal lines reproduced each 1/30th of a second (30 frames per second times 525 lines per frame equals 15,750 horizontal lines per second). This means the time for the trace of a line and its return to start another line is 1/15,750th of a second.

We should introduce the idea of talking about these extremely short time intervals in multiples of one millionth of a second. The unit of measurement is a microsecond. It is the length of time required for the completion of one cycle of carrier wave at the middle of the broadcast band, or 1000 kHz. The entire horizontal action, including the tracing of the picture line and the return to start a new line, occurs in 63.5 microseconds.

Let us divide the picture width (14 inches) by the time of active scanning (53.34 microseconds). We obtain a velocity of 4.1 miles per second. The retrace time (between points D and E in Fig. 7-1) is 7 microseconds. Since this retrace is over the same 14 inches of horizontal motion, the speed of

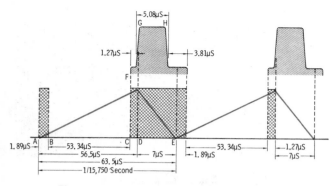

Fig. 7-1. Horizontal scanning waveform and synchronizing signal.

the spot (blanked out to produce no light) must obviously be much faster. Actually, this retrace can reach a speed of 31.5 miles per second.

As we previously stated, the sequence of events must occur exactly in step with a similar sequence occurring at the same instant in the camera tube at the transmitter. To accomplish this action, pulses are sent out from the transmitter between each horizontal trace. The shape of these pulses is shown above the sawtooth waveform in Fig. 7-1. At the instant shown as F, enough voltage appears at the grid of the picture tube to blank out all light. The region from F to G is known in

television slang as the "front porch." This region is slightly more than one-millionth of a second in duration. At point G, the carrier wave of the transmitter abruptly increases by approximately 25% of its average value. This sharp rise in the carrier triggers the scanning generators in the receiver. The scanning generators produce the required sawtooth motion of the electron beam. Exactly how the pulse accomplishes this triggering will be described later.

The horizontal beam does not trace a line parallel with the top of the picture, but has a slight downward slope. This vertical motion is controlled by a scanning sawtooth which moves the scanning spot to the bottom of the image and then rapidly returns it to the top. The electron beam moves from the top to the bottom of the picture and back to the top in 1/60th of a second. It is easy to see that this vertical scanning is much slower than the horizontal line tracing action and requires 16,666 microseconds. Pulses are sent out between successive fields to lock in, or control as a slave, the vertical-scanning oscillator of the receiver. A cycle of the vertical-deflection sawtooth waveform, together with an enlarged section of that part of the waveform which occurs during blanking and retrace, is shown in Fig. 7-2.

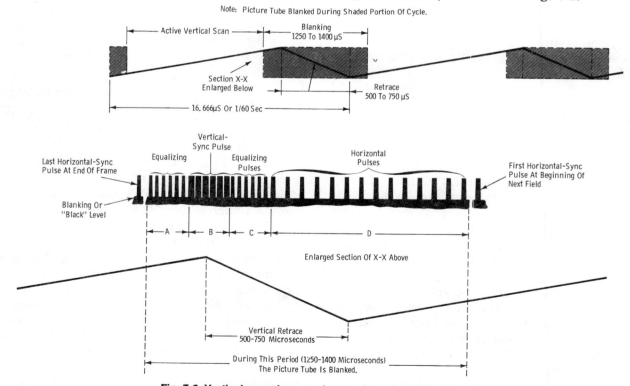

Fig. 7-2. Vertical scanning waveform and synchronizing signal.

The portion of the television signal which controls vertical retrace and synchronization is much more complicated than the single horizontal pulses which occur between successive horizontal lines. The vertical synchronizing signal resembles a comb with uneven teeth. If its only function were to trigger the vertical oscillator and to blank out the picture-tube screen during retrace, it could be made in the form of a single long rectangular pulse, the time duration of which would be from 20 to 22 horizontal lines (1250-1400 microseconds). However, the vertical synchronizing signal must perform two other functions. It must continue to keep the horizontal-scanning oscillator in step during vertical retrace and it must also assure that the alternate fields are properly interlaced.

Horizontal synchronization is kept in step by notches B and pulses A, C, and D (Fig. 7-2). Interlace is controlled by equalizing pulses A and C (Fig. 7-2) immediately preceding and following the vertical sync pulse.

We are not concerned at this time with the exact composition of the complex waveform making up the television signal; this subject is covered in Chapter 9.

CONTROL OF SCANNING GENERATORS BY SYNC PULSES

We have seen that the scanning systems of the receiver must keep in accurate step with the scanning raster of the camera tube at the transmitter. We have also described the type of synchronizing pulses which are made a part of the television signal to satisfy this requirement.

For a satisfactorily reproduced picture, the picture elements of adjacent horizontal traces must line up accurately, and the lines of alternate fields must interlace or space accurately between each other.

To avoid a displacement of more than one picture element in successive horizontal lines, the frequency stability of the horizontal oscillator must be 0.2% or better. Fig. 7-3A illustrates horizontal displacement.

To avoid "pairing" of the lines of successive fields (the lines lying on top of those of the preceding field instead of being properly interlaced), the stability of the vertical oscillator must be better than 0.05%. Fig. 7-3B illustrates this displacement.

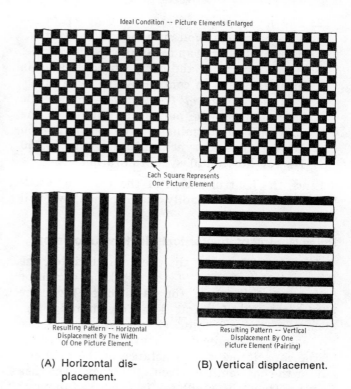

Ideal Condition -- Picture Elements Enlarged

Each Square Represents One Picture Element

Resulting Pattern -- Horizontal Displacement By The Width Of One Picture Element.

Resulting Pattern -- Vertical Displacement By One Picture Element (Pairing)

(A) Horizontal displacement.

(B) Vertical displacement.

Fig. 7-3. Picture-element displacement which might result from scanning-oscillator instability.

In each of the impulse-generating circuits (horizontal and vertical oscillator) suitable for television scanning, a coupling means is required in the input circuits for the introduction of synchronizing control pulses.

The horizontal and vertical pulses are clipped from the signal, amplified, and passed through circuits which classify the pulses so that each will control its respective scanning oscillator only. How these operations are accomplished will be described later. The end result is a short, sharp "pip" for the control of the horizontal oscillator and a long triangularly shaped pulse for control of the vertical oscillator.

In considering how the pulse controls the oscillator frequency, three items are important:

1. *The free-running frequency of the sweep generator.*—This frequency is that which would be generated at any particular setting of the hold control if the sync pulses were not present. The frequency can be slower or faster than the synchronizing pulse repetition rate, or in exact step. We will show later that, for proper stable operation, the slow rate is required.

2. *The firing point of the sweep generator.*—This is the bias voltage required to start capacitor discharge and scanning-wave retrace. At this point in the cycle, the oscillator is most sensitive to control by the sync pulse.

3. *The synchronizing frequency.*—This is the rate at which the pulses are applied to the control-input terminal of the oscillator—60 Hz for vertical synchronization and 15,750 Hz for horizontal synchronization.

Since the control action of the blocking oscillator can be more readily diagrammed, we will consider it first.

Pulse Control of the Vertical Blocking Oscillator

As we mentioned in the preceding chapter, the triggering, or firing, of the blocking oscillator occurs when the grid (or base-emitter) voltage passes the cutoff point. The free-running frequency of the oscillator (if no sync pulses are present) is determined solely by the time constant of the capacitance and resistance in the grid circuit or the base-emitter circuit, whichever the case may be. Once the oscillator is fired, it functions on its own until it is blocked again by the cutoff voltage.

If a pulse of positive potential from an external source is fed to the grid while the grid capacitor is discharging through the resistor, the grid volt-

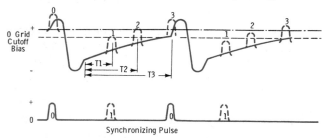

(A) Grid-voltage waveform with a series of sync pulses.

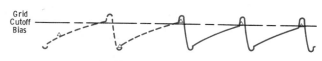

(B) Oscillator frequency fast.

(C) Oscillator frequency slow.

Fig. 7-4. Pulse control of free-running oscillator.

age will pass the cutoff point, and the tube will begin to conduct. The sawtooth-forming capacitor will discharge, and retrace will occur. A new scanning cycle then begins. Therefore, the repetition of the positive sync pulses can control the firing of the blocking oscillator and lock the picture into synchronization.

Fig. 7-4 illustrates this action in detail. Fig. 7-4A is an enlarged portion of the blocking-oscillator grid voltage. The synchronizing pulses below the grid waveform show a series of pulses marked "0," the leading edges of which are exactly in step with the grid-voltage waveform. These pulses do not affect the free-running frequency of the oscillator. They merely add to the grid voltage at the same instant it is being driven positive by the plate-current pulse. On the other hand, if the sync pulses occur at the points indicated as "1," the pulse voltage added to the discharge voltage of C1 through R1 (Fig. 6-11) is still short of the cutoff bias point and will not fire the tube. However, if the pulse occurs at points 2 or 3, the critical bias will be exceeded, the tube will immediately conduct, and retrace will begin.

Since the free-running frequency of the blocking oscillator can be changed by varying the time constant of the capacitance and resistance in the grid circuit, let us examine this action under the following conditions: (1) oscillator running faster than the sync-pulse rate, and (2) oscillator running slower than the sync-pulse rate.

Fig. 7-4B shows what happens when the oscillator is running faster than the sync-pulse rate. The dotted portion of the waveform indicates lack of synchronization at that point. Notice that several cycles occur before the pulse reaches point "X." At this point the grid cutoff voltage is exceeded, and the tube fires. Normally, you would expect the picture to lock in satisfactorily. However, this is not true. Lock-in occurs only momentarily during the field initiated at point "X." Succeeding fields do not lock in. With the oscillator running in this fast condition, the sync pulses are occurring during the scanning interval. Consequently, the picture is divided by the blanking bar. Also, an oscillator, running faster than the sync-pulse rate, can easily be triggered into erratic operation by automobile ignition and static interference. Therefore, the picture will be unstable. In modern receivers, however, improved circuit designs have resulted in more stable pictures, even under heavy interference conditions.

Fig. 7-4C shows what happens when the free-running frequency of the blocking oscillator is slower than the sync-pulse rate. Notice that lock-in occurs much faster and a good stable picture is obtained. This is obviously the desired condition because the oscillator should run slightly slower than the sync-pulse rate. The sync pulses will then take over and force the blocking oscillator to lock in with each succeeding sync pulse.

The height and width of the sync pulses are not important, as long as they are high enough to drive the grid above the cutoff point. With the sync-pulse amplitude reasonably high, the hold control can be varied over a fairly wide range without loss of synchronization.

For control of the blocking oscillator just described, the sync pulses are positive. When we study pulse clipping and amplification, we will find that a sync pulse can be made either positive or negative with respect to ground (or chassis), depending on receiver design. Negative pulses are ideal for the cathode-coupled multivibrator.

Pulse Control of the Cathode-Coupled Multivibrator

From Fig. 6-10 and the text, we know that "tripping" or discharge of the cathode-coupled multivibrator is initiated by a negative-voltage pulse on the grid of the first tube. Once "tripping" begins, the grid immediately receives additional negative voltage due to the current through the cathode-bias resistor, which is common to both tubes. Although the control actions and principles just described for the blocking oscillator hold true, it is not feasible to show them in diagram form. As a matter of fact, the small step in the leading edge of the grid-voltage curve of Fig. 6-10A (indicating the sync-pulse contribution to the grid voltage) is really a matter of speculation, since the action is so rapid and so cumulative that it is impossible to tell where pulse control stops and the circuit takes over.

Summary of Scanning-Generator Pulse Control

1. A positive synchronizing pulse controls the frequency of a blocking oscillator.
2. A negative pulse controls the frequency of a cathode-coupled multivibrator.
3. The free-running frequency of the scanning oscillator should always be slightly less than the synchronizing-pulse repetition rate. The

hold, or frequency, control of the oscillator controls this action.

4. As the grid or base voltage of a pulse generator approaches the "trigger" point, the oscillator becomes increasingly sensitive to control by additional grid or base voltage. At this point, the scanning can be "tripped" by interference. Special circuit combinations have been devised which are controlled by the pattern of the pulses rather than by the individual pulses. Such a system is relatively insensitive to interference, which seldom has a regular pattern. Its operation will be described later in the course when specific circuits are covered.

SCANNING REQUIREMENTS FOR PICTURE TUBES

Chapter 2 explained the theory and mechanical arrangement of the horizontal- and vertical-deflection coils. It stated that a sawtooth wave of current through the coils can be made to produce the desired raster. In other words, the amount an electron beam is deflected in an electromagnetically deflected cathode-ray tube depends on the strength of the magnetic field produced by the deflecting coils. The magnetic field is proportional to the amount of current passing through the coils, and these fields cross the path of the electron beam within the neck of the tube.

We must supply a linear sawtooth of current through the coil so that the electron beam will trace the proper raster under the combined influence of the horizontal- and vertical-deflection

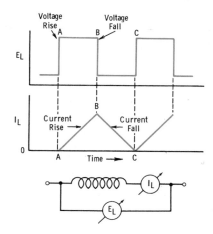

Fig. 7-5. Rise and fall of current through a pure inductance when a square-wave is applied.

coils. In Fig. 7-5 we see the resultant shape of a current wave through a pure inductance when a symmetrical square wave of voltage is applied across its terminals. This type of wave, as we have seen, can be developed by a conventional multivibrator. At point A, the voltage has suddenly been applied to the coil in much the same fashion as if a switch had been closed to connect the coil to a dc potential, such as a battery.

Notice that the current through the coil did not immediately rise to maximum. The self-induced voltage of the coil opposed the sudden change. The current, therefore, increased linearly over that portion of the cycle when the applied voltage was steady. (Theoretically, the current rises exponentially; but for practical purposes, we can consider it to be a linear change.) At point B, the impressed voltage was suddenly removed (the switch was opened). At this point, the current did not immediately fall to zero since it was maintained by the energy stored in the magnetic field. The self-induced voltage of the coil served as the driving potential to produce the linear fall of current from point B to point C.

We have now produced a triangular wave of current through the coil. If we can make the rise portion of the curve longer than the decay portion, we can produce the desired sawtooth-scanning current wave. This wave can be produced by making the impressed voltage wave asymmetrical, as shown in Fig. 7-6B.

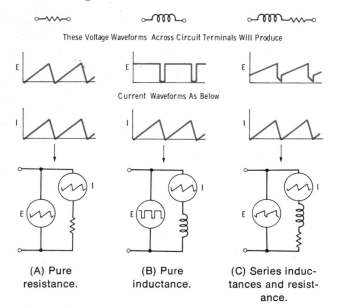

These Voltage Waveforms Across Circuit Terminals Will Produce

Current Waveforms As Below

(A) Pure resistance.

(B) Pure inductance.

(C) Series inductances and resistance.

Fig. 7-6. Voltage and current waveforms in inductance and resistance circuits.

Since a deflection coil cannot be built as a pure inductance, we must now consider what effect the resistance of the windings will have on the voltage waveform producing a sawtooth of current.

Fig. 7-6 illustrates three types of circuits and the voltage waveform necessary to produce a sawtooth wave of current through each circuit. Fig. 7-6A shows a pure resistance. The current is in phase with the voltage, and a sawtooth wave of voltage impressed across the resistor will cause a sawtooth wave of current through it. Energy losses occur only in the form of heat. The voltage required to produce a certain current is equal to the IR drop, as determined by Ohm's law.

Fig. 7-6C shows the circuit represented by a deflection coil. The voltage waveform will be seen as a combination of the sawtooth of A and the rectangular waveform of B (often called a trapezoidal waveform). In reality, this shape is the sum of an instantaneous pulse and a sawtooth. We can visualize its function as follows:

1. The sawtooth or linear rise portion of the waveform tends to produce a sawtooth waveform of current through the resistive part of the circuit.
2. The instantaneous pulse portion of the waveform forces a sawtooth waveform of current through the inductive part of the circuit.

To produce this combination waveshape, additional circuit elements are added to the sawtooth-capacitor charging circuit. The circuit is then known as a peaking type of waveshaping circuit. By proper choice of capacitor and resistor values, either the sawtooth portion or the impulse portion of the waveform can be made to predominate. The circuit action will be described later.

It is interesting to note that one part of the waveform must predominate over the other because of the differences between the horizontal- and vertical-deflection coils. In the vertical-deflection coil, the resistive component predominates over the inductive component. Thus, the sawtooth portion of the wave predominates over the impulse portion. For example, this coil might have a resistance of 68 ohms and an inductance of 50 millihenries. When the retrace rate is 60 Hz, as is the case for vertical sweep, a predominantly resistive circuit is presented.

In the horizontal-deflection coil of the same receiver, the conditions are reversed; the inductive component predominates. The impulse portion is

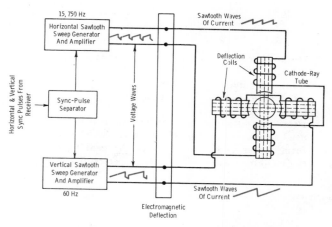

Fig. 7-7. Basic elements of an electromagnetic scanning system.

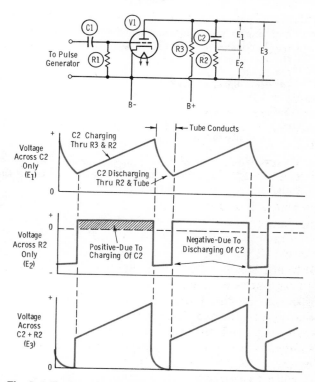

Fig. 7-8. Typical peaking circuit and associated waveforms.

more important, and the required waveshape approaches that of Fig. 7-6B. For example, we would find a resistance of only 14 ohms and an inductance of 8 millihenries. Since this coil operates at the much higher frequency of 15,750 Hz, the circuit is essentially inductive.

Fig. 7-7 shows, in block-diagram form, the basic elements of an electromagnetic scanning circuit.

PEAKING CIRCUITS FOR ELECTROMAGNETIC DEFLECTION

The combination sawtooth and impulse waveform required for electromagnetically deflected scanning can be generated by the peaking circuit shown in Fig. 7-8. The circuit action will be described in sequence.

1. The sawtooth-forming capacitor C2 is charged from the B+ source through resistors R3 and R2. This charging action takes place when the tube is not conducting.
2. The output-voltage waveform of the circuit is taken across the series combination of R2 and C2. R2 is known as the peaking resistor.
3. During the charging portion of the cycle, the voltage across the capacitor is a sawtooth wave.
4. When the tube conducts because of a positive pulse on its grid, the voltage across C2 and R2 is shunted by the low plate resistance of the tube.
5. The voltage across the capacitor cannot change instantly because the resistance of its discharge path through R2 and through the plate resistance of the tube is not zero.

Therefore, the difference in voltage must suddenly appear across peaking resistor R2. After this initial sudden change of voltage, the capacitor discharges exponentially through R2 and the tube until the tube again becomes nonconductive.

6. As the tube is cut off, the B+ potential is applied to the capacitor through R2 and R3 in series. Again, the capacitor voltage cannot rise instantaneously. The voltage across R2 must once more change abruptly, after which the capacitor charges through R2 and R3 in its normal sawtooth fashion.

By changing the values of R2 and C2, the ratio of the amplitude of the peaking impulse to the sawtooth can be adjusted to match the inductive and resistive requirements of the particular deflection coil.

QUESTIONS

1. How long does the beam take to move from the top to the bottom of the picture and back to the top again?

2. How many horizontal lines are produced per second?

3. What are the functions of the vertical-synchronizing signal?

4. Which one of the following pulses controls interlace:
   (a) Vertical-sync pulses?
   (b) Horizontal-sync pulses?
   (c) Equalizing pulses?
   (d) Horizontal pulses?

5. What is the polarity of the synchronizing pulse that controls the frequency of a blocking oscillator? Of a cathode-coupled multivibrator?

6. What is the current waveform passed through the deflection coils? What is the voltage waveform applied to the coils?

EXERCISES

1. Draw the current and voltage waveforms for the following:
   (a) Pure inductive circuit.
   (b) A series resistive and inductive circuit.

2. Draw the output stage of a sawtooth generator that is used to provide sweep voltage. Sketch the voltage waveform across each component of the peaking circuit and the combined waveform across the circuit.

CHAPTER 8

Deflection Systems

As discussed previously, the primary function of the deflection circuits in a television receiver is to move the electron beam in a manner that will produce the raster on the face of the picture tube. The vertical-deflection circuit moves the beam from the top of the screen to the bottom and then back to the top during retrace. The horizontal-deflection circuit produces beam movement from left to right and then back to the left side during retrace.

In most television receivers, another function of the horizontal-deflection circuit is to produce the high voltage which must be applied to the anode of the picture tube. In this chapter, we will discuss the various circuits used for deflecting the electron beam and producing the raster. We will also show how the high voltage is produced by the horizontal-sweep circuits. Fig. 8-1 shows the deflection circuits and high-voltage circuit in a tube-type, black-and-white television receiver. While this tube-type circuit is no longer used in television receivers now being manufactured, it is representative of the circuits which will be encountered in day-to-day servicing.

HORIZONTAL-DEFLECTION CIRCUITS

Fig. 8-2 shows the horizontal-deflection circuit used in a typical tube-type television receiver. This is a flyback system which utilizes the collapse of magnetic energy in the horizontal-output transformer and the horizontal-deflection yoke to complete the horizontal sweep and retrace action. When the horizontal-output tube is conducting, the current through the deflection coils causes the electron beam to move from the center of the screen to the extreme right. The tube is then cut off when the negative drive pulse arrives at the grid and the magnetic field in the deflection yoke collapses, causing the beam to move toward the left side of the screen.

We pointed out in Chapter 7 that horizontal retrace must be accomplished in the extremely short time of 7 microseconds. Because the horizontal-deflection coil system is predominantly inductive, a different method of operation from that of the lower-frequency vertical system must be employed. To obtain the rapid reversal of current through the horizontal-deflection coils, the output transformer and deflection coil circuit are tuned to a frequency of approximately 71 kHz by the associated circuit capacitances. This frequency is used because one-half cycle of oscillation is equal to the required retrace time of 7 microseconds.

The current through the deflection coils is maximum at either the extreme left or extreme right of the picture frame, with the axis (zero point) at the center. When the right-hand end of the trace is reached, the horizontal output tube is conducting high plate current, and a maximum of magnetic energy is stored in the deflection coils. At this instant, a negative pulse arrives at the grid of the horizontal-output tube and the output-tube plate current is suddenly cut off. The magnetic field in the transformer and in the deflection coils starts to collapse at a rate determined by the resonant frequency of the system (71 kHz). This collapse will shock-excite the circuit into damped oscillation, which, if allowed to continue, would produce the waveform shown in Fig. 8-3.

Continuation of this oscillation would seriously distort the left-hand side of the picture. An absorption, or damping, device is used to kill the oscillation during the 10 microseconds the picture

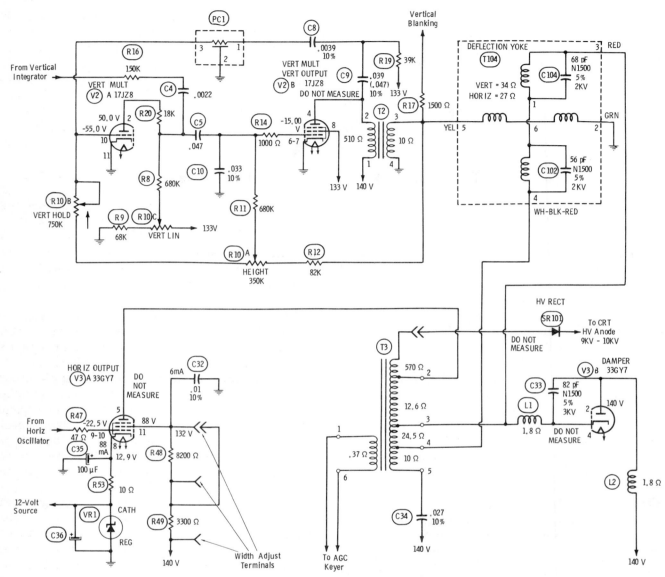

Fig. 8-1. Deflection circuits employed in a tube-type television receiver.

tube is blanked. This is the function of the damper tube in Fig. 8-2.

As the output-tube plate current is cut off by the negative scanning pulse, the induced voltage caused by the collapsing magnetic field prevents the damper tube from conducting, and no load will be imposed on the circuit. Therefore, the circuit oscillates for one-half cycle at its resonant frequency of 71 kHz (approximately 7 microseconds). This oscillation causes the current through the deflection coils to reverse to a maximum in the other direction and accomplishes the rapid return trace. The circuit would continue to oscillate for many cycles if it were not for the damper

tube, which now conducts during the second half-cycle.

As the current reverses through the deflection coils to start the second half-cycle of oscillation, the self-induced voltage also reverses. The damper tube then starts conducting and acts as a load across the deflection-coil system, and any further tendency toward oscillation is prevented. The current through the tube decays at a linear rate determined by the circuit constants. This linear current starts the next active scanning wave for the visible, or unblanked, trace.

If no additional current is supplied to the circuit from the horizontal-output tube, the electron

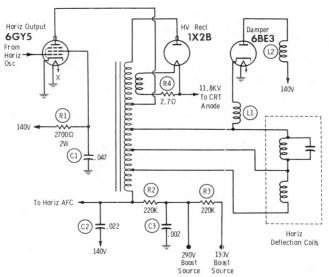

Fig. 8-2. Horizontal-deflection circuit used in a tube-type television receiver.

beam stops at the center of the picture tube as the current through the damper tube decreases to zero. The latter portion of the decay current wave departs from the desired linear sawtooth form. To overcome this nonlinearity, conduction of the horizontal-output tube is so timed that it starts contributing current to the deflection coils before the original current has completely decayed.

Fig. 8-3. Oscillation of horizontal-deflection-coil current if damper tube were not used.

This current contribution from the horizontal-output tube, which deflects the beam from the center to the right-hand side of the screen, is so shaped at its start that it corrects the nonlinearity of the original decay-current wave. As shown in Fig. 8-4, the two current waves overlap at the center of the scanning, and the combination produces a linear deflecting-coil current.

To summarize the circuit action to this point:

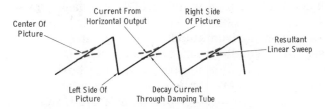

Fig. 8-4. Horizontal-scanning current waveforms.

1. The damper tube allows one-half cycle of natural resonant oscillation to occur in the deflection circuit, after which it loads the circuit and prevents further oscillation.
2. The first half-cycle of oscillation accomplishes beam retrace in the required 7 microseconds.
3. Decay of current through the damper produces the first half of the 53-microsecond active trace.
4. The horizontal-output tube starts to contribute power to the system before final decay of the damper-tube current and produces the final half of the active scanning wave.
5. At the end of the active scanning cycle, a negative pulse causes plate-current cutoff in the output tube and starts a new oscillation; the cycle is then repeated.

Width Control

The width of the raster (amount of horizontal deflection) was adjusted in some early sets by means of a variable inductor which shunted a portion of the secondary of the horizontal-output transformer. This shunting action controlled the output voltage and, hence, the width of the sweep. Since the width control also had a minor effect on the phase of the plate voltage of the output tube, it caused slight changes in the linearity of the right-hand side of the picture as the width was being changed.

In some sets, a metal sleeve around the neck of the picture tube is adjusted in and out of the deflection coil to vary the inductance and change the width of the horizontal sweep. Most late-model television receivers have no provision for width control.

Horizontal-Drive Control

Most tube-type black-and-white receivers have no provision for adjusting the horizontal drive. However, when such a control is provided, it may be connected as a variable resistance in the cathode circuit of the horizontal-output tube. Connected in this manner, the setting of the potentiometer controls the amount of input, or drive, applied between the grid and cathode of the output tube. The resistance of this control also protects the output tube by limiting the plate-cathode current to a safe value if the horizontal oscillator fails. The setting of this control determines the point on the trace at which the output tube con-

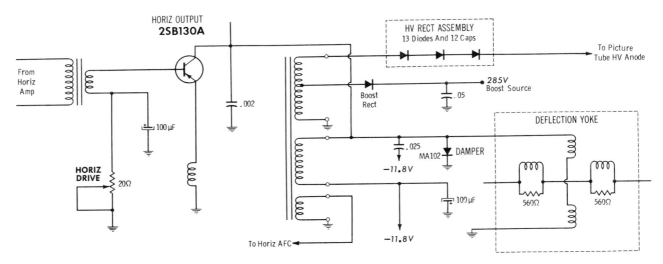

Fig. 8-5. Typical solid-state horizontal-deflection circuit.

ducts. Increasing its value widens the picture, crowds the right side, and stretches the left side.

In some early sets, the drive control was a trimmer capacitor in the grid circuit of the output tube. The coupling capacitor from the horizontal oscillator and the trimmer capacitor made up a voltage-divider network. By adjusting the trimmer capacitor, the drive signal applied to the grid of the horizontal-output tube could be increased or attenuated. Fig. 8-5 shows a solid-state horizontal-deflection circuit that has a horizontal-drive control.

Horizontal-Linearity Control

A horizontal-linearity control of some type was provided on practically all of the earlier receivers, but is seldom used in the black-and-white receivers of today. However, a horizontal-linearity control is found in some color receivers presently in use. This control is usually located in the cathode circuit of the damper tube, and its function is to permit small corrections in the shape of the saw-tooth current which is applied to the deflection coils.

Boosted B+

An important by-product of damper-tube action in tube-type horizontal-deflection systems is an increased voltage known as boosted B+ (or simply "boost voltage"). This voltage is derived from the inductive kickback produced by the sudden cutoff of output-tube plate current during the horizontal-retrace interval. The voltage provided by damper-tube conduction during this period is stored in the "boost" capacitors (C2 and C3 in Fig. 8-2) and is released when the damper tube is cut off after one half cycle of 71-kHz oscillation. This additional voltage is effectively added in series with the normal B+ supply voltage, "boosting" it to provide a much higher voltage for the plate of the horizontal-output tube and certain other stages in the receiver. The vertical oscillator and output stages, for example, are often supplied from this voltage source as is the focusing element of the picture tube. The boost voltage is often more than double the normal B+ supply voltage.

Solid-State Horizontal-Deflection Circuits

Fig. 8-5 shows the horizontal-output and high-voltage circuit used in a modern transistor tv receiver. The operation of this circuit is very similar to the tube-type circuit shown in Fig. 8-2. The use of a solid-state high-voltage rectifier assembly has eliminated the need for a high-voltage rectifier filament winding. The damper diode is connected across the horizontal-deflection coils and functions in the same manner as the tube-type damper shown in Fig. 8-2. The boost voltage is obtained by rectifying a pulse taken from a tap on the high-voltage winding of the flyback transformer.

Another horizontal-deflection circuit, which is used in a small-screen transistor set, is shown in Fig. 8-6. In this arrangement, the horizontal-deflection coils are connected directly to the emitter of the horizontal-output transistor. However, the horizontal-output transformer functions in the

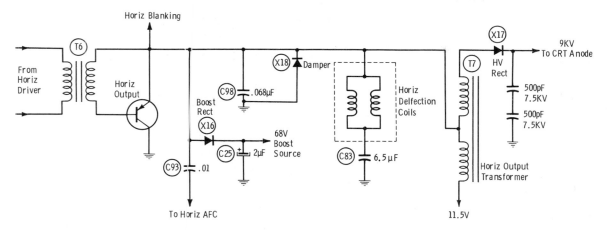

Fig. 8-6. Another example of a solid-state horizontal-deflection circuit.

same manner as the horizontal-output transformers that we have discussed previously. The deflection coils are effectively connected across the primary of the output transformer which forms part of the 71-kHz resonant circuit.

Note that the horizontal-deflection coils are connected in parallel. This is done to provide the low-impedance load required for transistor circuits. Since the inductances of the horizontal-output transformer and deflection coils used in solid-state receivers are somewhat lower than the inductances found in tube-type sets, the capacitances in the circuit must be larger in order for the circuit to be resonant at 71 kHz.

Regardless of whether tubes or transistors are used in the horizontal-deflection circuit, the circuit operates in the same manner. The horizontal-output stage is conducting during that portion of the forward trace which moves the electron beam from the center of the screen to the right side. When the beam reaches the extreme right side of the screen, the horizontal-output stage is cut off. This causes the horizontal-deflection coils, the horizontal-output transformer, and the associated circuit capacitances to oscillate at their resonant frequency (71 kHz). During the first half cycle of oscillation, retrace occurs and the beam is returned to the extreme left side of the screen. At the end of the first half cycle of oscillation, the voltage polarity across the damper is such that it conducts, causing the oscillation to stop. The current through the deflection coils during the time that the damper conducts causes the electron beam to move from the left side of the screen to the center. When the beam reaches the center of the screen, the damper stops conducting, and the hor-

izontal-output stage comes out of cutoff. The cycle just described starts over again, and forward trace is completed.

HIGH-VOLTAGE POWER SUPPLY

An important function of the horizontal-output circuit is to supply the anode voltage for the picture tube. The high-voltage supply differs considerably from the low-voltage supplies discussed in Chapter 4. The current is quite small, usually around 300 microamperes, while the voltages are quite high. Present-day receivers may have cathode-ray tube accelerating potentials from 5000 to 30,000 volts.

Since high-voltage power supplies represent extremely dangerous shock hazards, it is wise to follow normal precautions when working with them. When making high-voltage measurements, it is a good idea to use only one hand and keep the other hand away from the chassis.

Flyback High-Voltage System

The flyback method of obtaining high voltage is used in nearly all present-day television receivers. It utilizes the high-voltage pulse created in the primary of the horizontal-output transformer during retrace. With this system, few additional components are required, since practically all magnetically deflected receivers use a matching transformer (horizontal-output transformer) between the horizontal-output stage and the deflection coils. This system also prevents modulation of the video signal by stray energy from the high-voltage supply, since the picture is blanked during retrace. In the circuit of Fig. 8-6, the blanking pulse is taken from the emitter of the horizontal-output transistor.

The addition of two windings to the horizontal-output transformer in a tube-type receiver makes it possible to produce the high voltage needed at the anode of the picture tube. When the current through the horizontal-output stage collapses due to the sawtooth voltage applied to the grid or base, a relatively high pulse voltage will be produced across the primary of the output transformer because of self-induction. By the addition of an auto-transformer secondary winding to the horizontal-output transformer, the pulse voltage can be stepped up to almost any desired value. The high-voltage pulse is then fed to the plate of a rectifier where it is rectified and filtered to produce the high dc potential for the anode of the picture tube.

A tertiary winding consisting of one or two turns on the transformer provides the filament voltage for the high-voltage rectifier. This is possible since a relatively low filament current is required for this tube. When a solid-state high-voltage rectifier is employed, the filament winding is not necessary, of course.

Because of the high horizontal-scanning frequency (15,750 Hz), a powdered-iron core can be used for the horizontal-output transformer. The windings are of the universal type and are usually impregnated with wax. Since the frequency is high, and the current drain is low, the filtering requirement is small. In modern sets, filtering is accomplished by using the capacitance between the inner and outer layers of aquadag coating on the picture tube. A typical horizontal-output transformer is shown in Fig. 8-7.

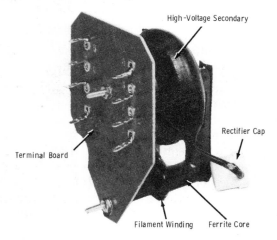

Fig. 8-7. Typical horizontal-flyback transformer used in a tube-type set.

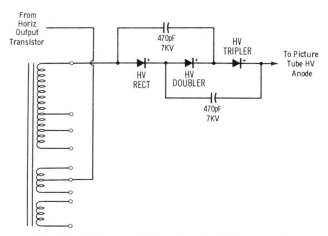

Fig. 8-8. Solid-state high-voltage multiplier circuit.

High-Voltage Multiplication

In the systems previously described, a single rectifier was employed to obtain the high voltage required for the picture tube. With the first solid-state receivers, however, the voltage and current limitations of the rectifying components were such that voltage multiplication was required in the high-voltage circuit. Fig. 8-8 shows a voltage tripler configuration employed in some transistor television receivers. The majority of early solid-state receivers used subminiature tubes in the high-voltage rectifier circuit, as shown in Fig. 8-9. Filament voltages are obtained from three separate windings on the horizontal-output transformer.

In the circuit of Fig. 8-9, tubes V1 and V3 are driven into conduction on the large positive flyback pulses, and V2 conducts on the negative swings. The negative swings are coupled to the filament of V2 through capacitor C2. All three capacitors (C1, C2, and C3) are kept charged at a level that results in a dc output voltage greater than twice the value of the ac input voltage.

The development of solid-state rectifiers with higher current and voltage ratings has eliminated the use of vacuum-tube high-voltage rectifiers in transistor receivers now being made. When solid-state rectifiers are used in lieu of tubes, the circuit is simplified by the absence of filament windings.

HORIZONTAL-OSCILLATOR CIRCUITS

The function of the horizontal-oscillator circuit is to determine the conduction and cutoff periods of the horizontal-output stage. The oscillator must be exactly in step with the horizontal-scanning

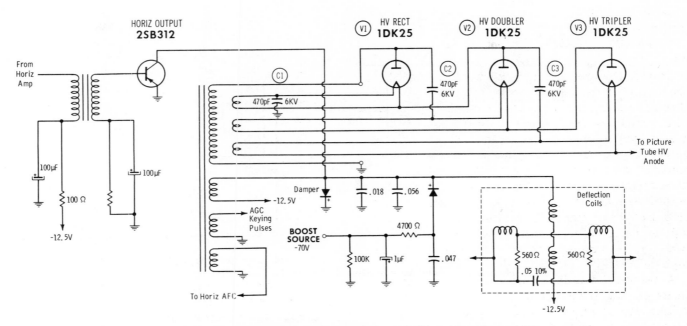

Fig. 8-9. Example of vacuum-tube high-voltage multiplier circuit used in early transistor receivers.

generator at the transmitter in order for the picture to be properly synchronized. The sync pulses discussed in Chapter 7 are used to control the oscillator and keep it in step with the transmitted picture. The processing of these sync pulses will be discussed in Chapter 10. Here, we are concerned with the operation of the oscillator circuit itself.

Cathode-Coupled Multivibrator

The cathode-coupled multivibrator is the horizontal-oscillator circuit employed in most present-

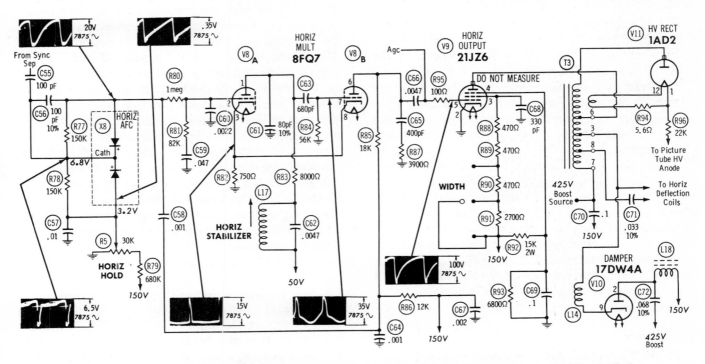

Fig. 8-10. Example of a horizontal-deflection system using a cathode-coupled multivibrator controlled by afc diodes.

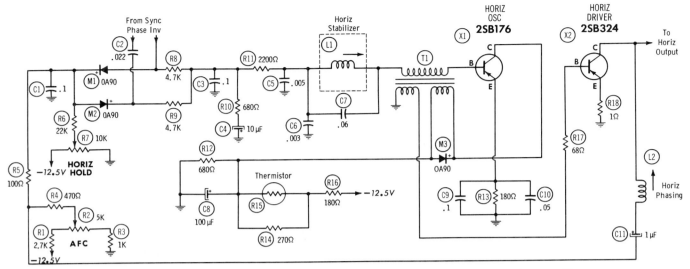

Fig. 8-11. Partial schematic of a solid-state horizontal-deflection system using a blocking oscillator.

day tube-type receivers. Fig. 8-10 shows a typical horizontal deflection system using a cathode-coupled multivibrator. The waveforms at various points within the circuit are also shown. In this circuit, the free-running frequency of the multivibrator is controlled by the setting of hold control R5 in the horizontal afc (automatic frequency control) circuit.

A sample of the scanning-wave voltage is coupled through C58 back to afc diode assembly X8 where it is compared with the repetition rate of the horizontal-sync pulses received from the sync separator. The afc circuit produces a dc voltage, which represents the phase difference between the two signals, and this voltage is applied to the grid of the multivibrator in order to control its firing point. Thus, the afc circuit monitors the frequency of the horizontal-oscillator circuit and automatically corrects for any variation from the horizontal-sync rate.

The setting of horizontal-hold control R5 determines the control range of the horizontal afc circuit. The tuned circuit consisting of L17 and C62 in the plate circuit of the first section of the multivibrator is referred to as the horizontal stabilizer. Adjustment of this circuit provides additional control of the free-running frequency of the multivibrator. This is a service adjustment used to set the range of the hold control, so that it operates symmetrically about its midpoint.

Blocking Oscillator

Although the blocking oscillator is not used as the horizontal oscillator in modern tube-type re-

ceivers, it is a popular configuration in solid-state sets. Fig. 8-11 shows a typical solid-state configuration consisting of the horizontal-afc, blocking-oscillator, and horizontal-driver circuits.

The afc circuit (consisting of diodes M1 and M2 and their associated components) in this case is the series type which is driven by push-pull horizontal sync pulses from a sync-phase inverter. Comparison pulses are fed to the junction of the two diodes from the collector of driver transistor X2. The base bias for the oscillator transistor X1 depends on the settings of R2 and R7 and the net value of the reference voltage produced by M1 and M2. A negative sync pulse is applied to the cathode of diode M2, and a positive sync pulse is fed to the anode of diode M1. The net value of the voltage produced at the junction of R8 and R9 depends on the relative conduction of diodes M1 and M2. This conduction is determined by the combined peak voltages of the sync pulses and the phase of the comparison pulse. When the phase of the comparison pulse shifts away from the phase of the sync pulses, the rectified output of the afc diodes at the junction of R8 and R9 then changes. This positive or negative voltage change at the base of transistor X1 keeps the horizontal blocking oscillator "on frequency."

The RC network composed of C3, R10, C4, R11, C5, and C6 serves as a filter for the afc control voltage and also prevents out-of-phase feedback which could cause motorboating. Coil L2 and capacitor C11 in the feedback circuit provide the proper waveshape for the drive applied to the base of the blocking oscillator. Circuit oscillation is

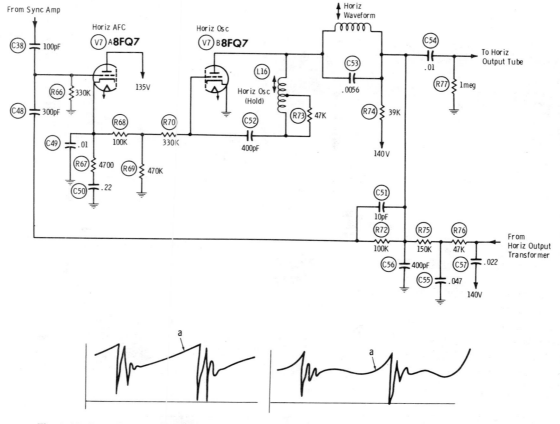

Fig. 8-12. A modern pulse-width circuit, and waveform showing the effect of the stabilizing coil.

achieved by positive feedback between the collector and base of X1 through blocking transformer T1. Diode M3 short-circuits the overshoot pulse at the trailing edge of the collector waveform and thereby provides waveshaping, as well as protection, for oscillator transistor X1.

Pulse-Width Systems

A horizontal-oscillator circuit, which utilizes the pulse-width system for controlling the frequency of the oscillator, is shown in Fig. 8-12. This type of circuit was used extensively in later tube-type receivers. A pulse from the sync amplifier is applied to the grid of the control tube along with feedback from the output of the oscillator and a reference pulse from the horizontal-output transformer. Part of the cathode circuit of the control tube is common to the grid circuit of the oscillator. Thus, the control tube can affect the oscillator frequency by influencing the time constant of the oscillator grid circuit which includes R67.

A sharp pulse voltage derived from the horizontal-output transformer is combined with a small portion of the sawtooth voltage from the plate circuit of the horizontal oscillator. This combination produces a sawtooth voltage which has a very sharp trailing edge and is applied to the grid of the control tube. It will coincide with the original sync pulse only if the oscillator is in exact step with the sync pulses.

As exact synchronization is reached, the control-tube grid pulse, which consists of the original sync pulse added to the feedback pulse, will be narrow and of high amplitude. If the feedback pulse is slightly fast or slow, it will not add to the original sync pulse but, instead, will widen the original pulse. From this combination of variable height and width of the voltage pulse on the grid of the control tube, very precise timing of the horizontal oscillator is achieved. The plate-current pulse of the control tube through cathode resistor R67 and capacitor C50 adjusts the grid-circuit time constant of the oscillator and produces the required exact synchronization.

A horizontal-hold adjustment is provided for the circuit of Fig. 8-12 by adjusting the slug of

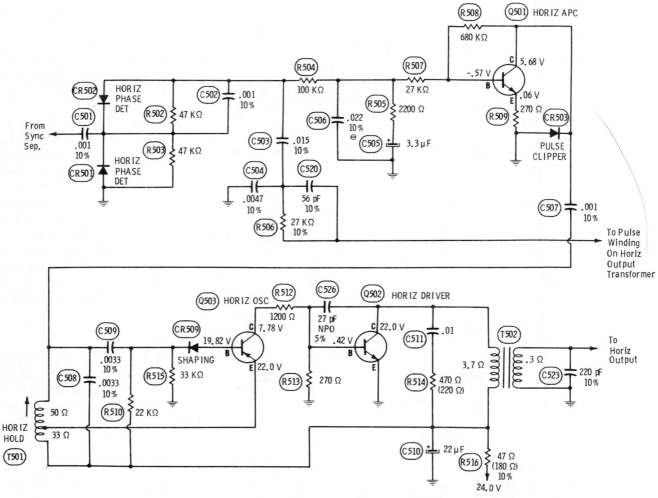

Fig. 8-13. Solid-state horizontal-oscillator circuit using a modified Hartley oscillator.

L16. A further refinement, which has been used with various pulse-width systems in the past, is the horizontal-waveform coil, L17. This coil is also called a horizontal stabilizing coil. This inductance and the capacitance across it are tuned to the horizontal-scanning frequency and will produce a sine wave of voltage in series with the plate circuit. The sine-wave voltage steepens the slope of the grid-voltage waveform immediately before conduction time. This action improves the frequency stability of the oscillator by reducing the possibility of random noise triggering the oscillator. Waveforms A and B in Fig. 8-12 show the oscillator grid voltage, and point a in each waveform shows the change of slope.

Hartley Oscillator

The horizontal oscillator circuit in Fig. 8-13 operates as a modified Hartley oscillator consisting of Q503, T501, and capacitor C508. The dc supply voltage for the emitter of Q503 is supplied through the lower half of T501 and capacitor C509 is used to provide dc blocking at the base of Q503. The 15,750-Hz signal developed by the horizontal oscillator is applied to the base of horizontal-driver transistor Q502 through resistors R512 and R513. These resistors provide both ac coupling and dc bias to the base of Q502. This bias voltage causes the driver transistor to operate as a switch, going from saturation to cutoff at the rate of the horizontal oscillator. In turn, this action produces the correct horizontal-drive waveform at the secondary winding of the driver transformer. This drive signal is then applied to the base of the horizontal output transformer.

The afc circuit functions in the following manner. A negative horizontal sync pulse from the sync separator is applied through capacitor C501 to the cathodes of horizontal afc diodes CR501 and CR502. A negative gating pulse from a winding on the horizontal output transformer is developed into a sawtooth waveform by resistor R506 and capacitor C520. This sawtooth waveform is coupled to the anode of afc diode CR502 through capacitor C503. If the horizontal oscillator and the sync pulses are in phase, no error voltage will be developed by the afc circuit. However, when the horizontal oscillator frequency leads or lags the horizontal sync pulses, a positive or negative error voltage will be developed at the anode of afc diode CR502. This error voltage will be algebraically added to the positive voltage at the base of afc transistor Q501. As a result, the effective resistance of Q501 is changed. This varying resistance, which is in series with capacitor C507, corrects the horizontal oscillator and keeps it in phase with the horizontal sync pulses. The afc antihunt network, which consists of capacitors C505 and C506 and resistor R505, provides stable operation of the circuit and avoids self-oscillation.

VERTICAL-DEFLECTION SYSTEMS

As discussed previously, the function of the vertical-deflection system is to move the electron beam from the top of the screen to the bottom and then back to the top during retrace. This cycle must be repeated 60 times per second, which is the vertical-sweep rate. The vertical-sweep circuits produce the sawtooth current, which must be applied to the vertical-deflection coils in order to provide proper vertical deflection.

The vertical-deflection system is generally less complex than horizontal-deflection systems and, therefore, involves fewer components. The system usually consists of the oscillator section, an output stage, and their associated circuits. In solid-state sets, an amplifier stage is sometimes used between the oscillator and the output stage. The circuits most often employed as vertical oscillators are the multivibrator and the blocking oscillator. In newer tube-type sets, which usually use the multivibrator as the vertical oscillator, the second stage of the multivibrator often does double duty as the vertical output stage.

Vertical Blocking Oscillator

The vertical blocking oscillator was quite popular in early television receivers but was later all but abandoned in favor of the multivibrator circuit in tube-type sets. However, since the introduction of transistor television receivers, the blocking

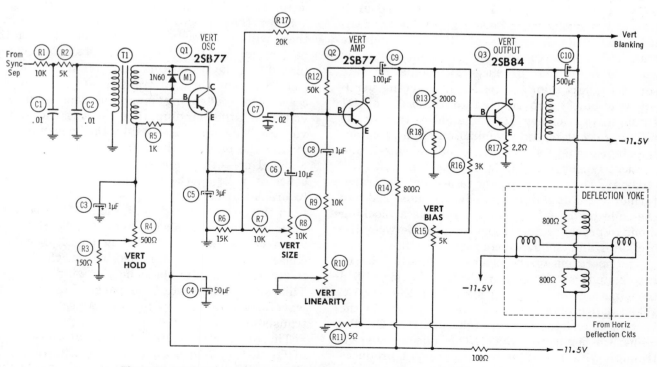

Fig. 8-14. Typical transistor vertical-sweep system using a blocking oscillator.

oscillator is once again quite popular in vertical deflection systems.

Fig. 8-14 shows a typical example of a solid-state vertical-deflection system using a blocking oscillator. The majority of solid-state receivers employ three stages in the vertical sweep system. These consist of the vertical oscillator, the vertical amplifier (also sometimes referred to as the vertical driver), and the vertical-output stage.

In the blocking oscillator circuit of Fig. 8-14, transistor Q1 operates in class-C, conducting only for brief intervals. During conduction, vertical-oscillator transistor Q1 generates pulse waveforms. The base is driven strongly by the amplified output of the collector during conduction by feedback through the tightly coupled windings of transformer T1. As the oscillator transistor conducts, the base-emitter junction rectifies the negative-going pulses and charges C5 to a relatively high negative dc potential. At the same time, transistor Q1 saturates and there is no further increase in collector current. Since the collector current is now constant, the base is no longer driven by feedback through transformer T1 and the transistor is cut off, or blocked.

After the transistor is cut off, capacitor C5 then begins to discharge through resistor R6. When the capacitor has discharged sufficiently, transistor Q1 begins to conduct, and C5 is again charged until conduction is blocked. The setting of the vertical-hold control R4 determines the time at which the transistor begins to conduct. The hold control is set to a point that produces an oscillation rate somewhat less than 60 Hz. Negative-going vertical-sync pulses are coupled to the base of Q1 through transformer T1. These pulses trigger the blocking oscillator into conduction slightly before the stage would normally come out of cutoff, and vertical synchronization is thereby achieved.

Diode M1 acts as a limiter to prevent the transformer from applying excessive voltage peaks to the collector of Q1 and also serves as a blocking diode to prevent undesirable coupling between the oscillator and the sync-separator circuit.

The operation of the blocking oscillator can be better understood by considering the simplified blocking-oscillator circuit shown in Fig. 8-15A. When a negative trigger pulse is applied to the base of transistor Q1 through capacitor C1, the transistor begins to conduct. The collector current through winding 1-2 of transformer T1 produces a magnetic flux that induces a voltage of opposite

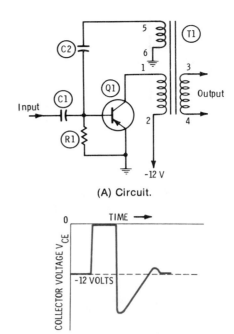

(A) Circuit.

(B) Collector-voltage waveform.

Fig. 8-15. Simplified blocking-oscillator circuit and collector-voltage waveform.

polarity in winding 5-6. This voltage is coupled through capacitor C2 to the base of Q1 in order to provide regenerative feedback. This quickly drives the transistor into saturation and the collector current cannot rise further.

Since the collector current is now constant, there will be no feedback voltage induced into the base circuit of Q1. Therefore, the base is reverse-biased and the transistor is cut off. When the collector current through transformer winding 1-2 ceases, the collapsing magnetic field induces a voltage in the winding that exceeds the supply voltage. This is represented by the backswing following the pulse in the waveform shown in Fig. 8-15B. Unless the transistor is rated for this peak collector voltage, the collector junction is likely to break down.

In the circuit of Fig. 8-14, diode M1 is reverse-biased when transistor Q1 is conducting. However, when Q1 is cut off, the rising collector voltage causes M1 to become forward-biased which effectively short-circuits the collector winding of T1. This action eliminates the backswing from the waveform shown in Fig. 8-15B and protects the transistor from breakdown.

The pulse waveform developed at the emitter of Q1 in Fig. 8-14 is changed into a sawtooth by C5 and R6. These are the same components that

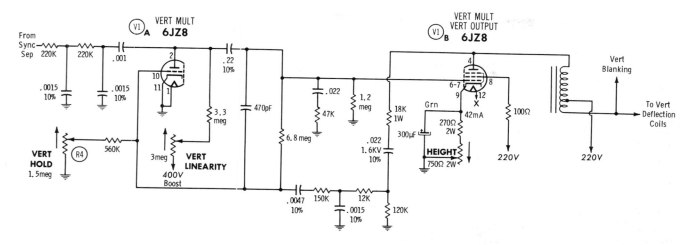

Fig. 8-16. Vertical-deflection circuit using a multivibrator.

provide the time constant for the oscillator circuit. The discharge of C5 through R6 produces a curved exponential waveform. Negative feedback through resistor R17 makes the curved waveform linear. Circuit tolerances are such that manual control of vertical linearity is necessary which is provided by vertical-linearity control R10.

Vertical-size control R8 is used to control the amplitude of the drive signal applied to the base of the vertical-amplifier transistor, Q2. This control makes it possible to adjust the height of the raster. The vertical-deflection coils are returned to ground through the emitter resistor for Q2. This provides negative feedback for the vertical-amplifier stage and improves the overall linearity of the system.

Since vertical-output transistor Q3 is operated near its maximum rated output, the collector junction heats up appreciably. In order to prevent thermal runaway and possible damage to the transistor, the base-emitter voltage of Q3 must be stabilized. When the base-emitter current of Q3 tends to increase, the temperature of thermistor R18 increases and its resistance decreases. This decreases the base voltage of Q3, thereby stabilizing its operating point. The base-emitter bias voltage for output transistor Q3 is adjusted manually by vertical-bias control R15.

Vertical Multivibrator

The multivibrator is by far the most popular vertical-oscillator circuit used in modern tube-type receivers. The vertical-deflection circuit of Fig. 8-16 employs an asymmetrical multivibrator and a vertical-output stage. This particular cir-

cuit consists of a 6JZ8 triode-pentode. The pentode section of this tube serves not only as the second stage of the multivibrator, but also as the vertical-output stage. As in the multivibrator circuit discussed previously, the free-running frequency is determined by the RC time constant which is made variable by vertical-hold control R4.

The vertical sweep circuit shown in Fig. 8-17 has been employed in many receivers and uses a special tube designed for this application. The tube consists of two dissimilar triode sections. The output section will provide more power amplification than the oscillator section.

Like all multivibrators, oscillation takes place due to the alternate conduction of the two sections of the tube. The circuit is free-running and will oscillate at its natural frequency without the necessity for external triggering. However, to synchronize the oscillation with a signal coming from the transmitter, positive-going sync pulses are coupled to the circuit through capacitor C1.

Tube V1A acts somewhat as a switch which automatically charges and discharges the sawtooth-forming capacitor C4. During vertical-trace time, V1A is not conducting, and capacitor C4 charges. The positive-going signal voltage developed across the combination of C4, C3, and R6 is then coupled to the grid of V1B, where it causes a sawtooth deflection current in the plate circuit and through the yoke.

When the trace portion of the vertical-scan cycle is completed, tube V1A conducts, and capacitor C4 discharges so that a sharp negative voltage is applied to the grid of V1B and cuts it off. The circuit is designed so that V1A conducts only during

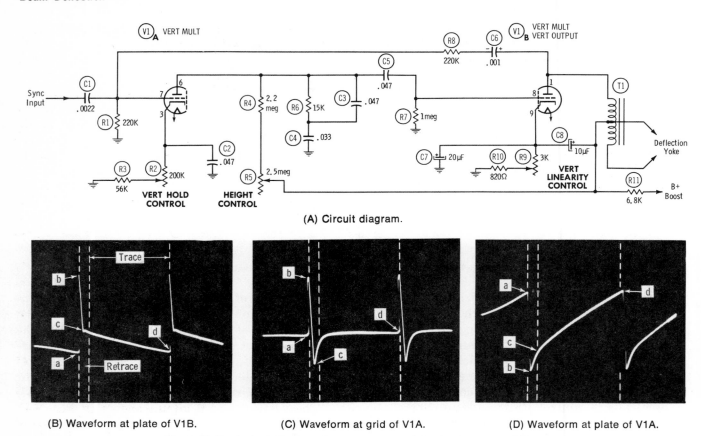

(A) Circuit diagram.

(B) Waveform at plate of V1B.

(C) Waveform at grid of V1A.

(D) Waveform at plate of V1A.

Fig. 8-17. Vertical oscillator and output system with associated waveforms.

the retrace period, or for approximately $\frac{1}{15}$ of the vertical-scan cycle.

The following detailed circuit description begins with the moment that the set is turned on. With the cathodes heated and plate voltage applied, both tubes will start to conduct. However, for the circuit to oscillate, one tube section must conduct while the other is cut off. Let us assume that this is initially accomplished by the action we will discuss next.

As the B+ voltage is applied to the circuit, capacitors C5 and C6 attempt to charge to their associated plate potentials. Because of the relatively low resistance in the charge path of C6 and the higher plate voltage on V1B, the initial capacitor-charging current through R1 and R8 will exceed that through R7, R4, and R5. The current through grid resistor R1 causes the grid voltage of V1A to rise slightly in a positive direction before the grid voltage of V1B rises. The positive-going voltage on the grid of V1A increases the plate current through the tube. More voltage is dropped across resistors R4 and R5, and the plate potential is lowered. Since this negative-going signal is cou-

pled to the grid of V1B through capacitor C5, the conduction of V1B is diminished, and its plate voltage increases. As the plate voltage of V1B rises, capacitor C6 couples the positive voltage back to the grid of V1A, which causes it to conduct still more. The variation in plate current between the two triode tube sections is amplified by this continued action until V1A is highly conductive and V1B is completely cut off. This is the action which occurs between points a and b on waveforms B, C, and D in Fig. 8-17.

Vertical-output transformer T1 represents a relatively large value of inductance in the plate circuit of V1B. When the output triode is cut off, current through T1 ceases, and the magnetic field produced by the current collapses. The voltage induced across the transformer by this collapsing field is very high and accounts for the high value (approximately 1400 volts) of the pulse at point b in the waveform of Fig. 8-17B. Since V1B is cut off, its plate voltage drops rapidly toward point c after the magnetic field has collapsed.

A portion of the high pulse voltage developed across the top section of T1 is divided between

R8 and R1. Note the resulting positive grid swing in the waveform of Fig. 8-17C. During this pulse, the grid of V1A conducts and charges C6 in the polarity shown in Fig. 8-17A. When the plate voltage on V1B drops to point c in Fig. 8-17B, the charge on capacitor C6 adds to this negative voltage swing and causes the grid voltage of V1A to drop all the way to point c in Fig. 8-17C. This negative grid voltage cuts off tube V1A.

To explain the shape of the waveform between points b and c in Fig. 8-17D, we must say that the charge on C4 and C3 is immediately dissipated when V1A conducts. Shortly after time b, when the grid voltage starts in a negative direction and the plate current begins to fall off, the plate voltage rises once again. However, by the time the tube cuts off, the voltage reaches point c only. From point c to point d, capacitors C4 and C3 slowly charge through the path consisting of the power supply, R11, R5, and R4. This action accounts for the sloping rise of the plate voltage of V1A (Fig. 8-17D) during this period.

The almost linear shape between points c and d of Fig. 8-17D produces a steady increase in the grid voltage of V1B and in the plate current. This is the reason for the steady decrease of V1B plate voltage between points c and d in Fig. 8-17B.

Tube V1A is now cut off and V1B is conducting; this is opposite of the initial circuit action. Conduction of V1A does not start immediately again after time c, because its grid voltage is held just a few volts more negative than the cutoff voltage. The level portion of the trace between points c and d in the waveform of Fig. 8-17C represents this bias level. The exponential rise begin-

ning at point c represents the decrease in the discharge current of C6. During the remainder of the trace period, C6 continues to discharge through R1 at a lesser but steady rate because of the linear decrease in V1B plate voltage. The steady rate of discharge through R1 holds V1A at cutoff during trace time. Another factor contributing to the maintenance of cutoff of V1A is the low plate voltage during the early part of the trace period.

All inductors have a saturation point, that is, a point at which additional current will not cause further expansion of the magnetic field. This is what happens within a blocking-oscillator transformer and, in this case, within output transformer T1. Current through V1B increases linearly until T1 becomes saturated. When this saturation occurs, the plate voltage of V1B will no longer decrease, and C6 will cease to discharge through R1. With the negative bias removed, V1A starts to conduct, and the whole cycle is repeated.

The frequency of oscillation in this particular circuit is determined by the charge and discharge of capacitors C4 and C3 and the setting of vertical hold control R2. The setting of the vertical hold control determines the bias on V1A and, thus, the time at which it starts to conduct. The vertical sync pulses, which are coupled to the grid of V1A by way of C1, tend to force the oscillator into sync with the incoming signal.

A solid-state vertical-sweep circuit using a multivibrator as the vertical oscillator is shown in Fig. 8-18. The operation of this multivibrator is similar to the transistor multivibrator circuit discussed in Chapter 6. The vertical-hold control (R1) is in the base circuit of multivibrator tran-

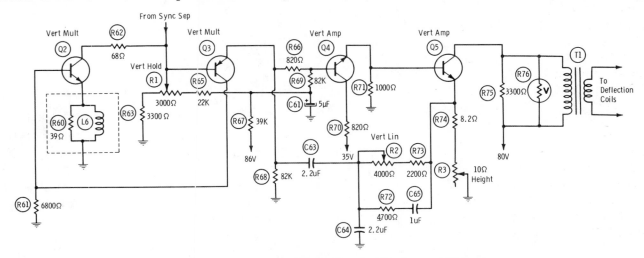

Fig. 8-18. Solid-state vertical-deflection circuit using a multivibrator.

sistor Q3. This control adjusts the bias on the transistor and determines the triggering time for the multivibrator. Two capacitors connected in series, C63 and C64, make up the sawtooth-forming capacitor. A small parabolic voltage from the emitter circuit of the vertical output transistor is fed to the junction of these two capacitors. This correction voltage is used to linearize the waveform produced by the sawtooth-forming circuit.

An additional stage, the vertical amplifier, is used between the multivibrator and the vertical output stage. Since this stage is connected as an emitter follower, it actually provides no gain. Its purpose is to provide impedance matching between the relatively high output impedance of the multivibrator and the rather low input impedance of the output stage. Vertical linearity control R2 varies the amount of feedback voltage applied to the sawtooth-forming circuit. Height control R3 is used to vary the emitter bias of the output transistor. The voltage-dependent resistor (R76) across the primary of the vertical-output transformer keeps the amplitude of the drive voltage applied to the vertical-deflection coils essentially constant. When the amplitude of the output waveform tends to increase, the resistance of R76 decreases.

QUESTIONS

1. What are the main purposes of the damper circuit?

2. Is the horizontal-deflection coil system predominantly resistive or inductive?

3. Does the damper tube conduct during the first or second half of the trace period?

4. What potentials can be expected at the output of the high-voltage supply?

5. What is created in the primary circuit of the horizontal-output transformer for use in developing the high voltage? During what scanning period is it developed?

6. What is the function of the horizontal afc circuit?

7. What is the purpose of the horizontal-stabilizer control in many horizontal-multivibrator circuits?

8. What is the purpose of the resistors connected across the vertical-deflection coils?

9. What type of network is used to reject the horizontal-sync pulses and to accept the vertical-sync pulses? What type of network accepts the horizontal-sync pulses?

EXERCISES

1. Show, by using a block diagram, the stages of a vertical-deflection system by starting at the output of the sync separator. Describe the function of each stage.

2. Show, by using a block diagram, the stages of a horizontal-deflection system by starting at the sync separator. Describe the function of each stage.

Beam Modulation and Synchronization

CHAPTER 9

The Composite Television Signal

In order to understand the operation of the remaining sections of the television receiver, we must study the nature of the composite television signal at this point.

The television signal is more complex than standard broadcast signals, which carry only audio information. To better present the reasons for the complexity of the television signal, we will review the composition of the familiar amplitude-modulated (am) broadcast carrier before describing the television carrier.

During the amplitude modulation of a carrier wave by speech or music, frequencies higher and lower than the carrier frequency are produced. These frequencies are known as sidebands. These sidebands occur as a result of the beat, or heterodyne, between the carrier frequency and the modulating frequency. In the am broadcast band, for example, the highest modulating frequency for speech or music transmission is approximately 5000 Hz. Thus, if a broadcast station operates at 1000 kHz (one million hertz), the 5000-Hz modulation will produce sidebands that extend from 1,000,000 Hz minus 5000 Hz to 1,000,000 Hz plus 5000 Hz (995 kHz to 10005 kHz). A double-sideband modulated broadcast carrier is shown in Fig. 9-1.

The modulating frequencies encountered in the television video signal extend from less than 30 Hz to over 4,000,000 Hz. It will be interesting to examine the reasons for this tremendous frequency range in the output of the transmitter camera tube.

As the signal spot in the camera tube sweeps across the scene, approximately 480 lines are scanned in 1/30th of a second. (Although there are 525 lines per frame, only 480 lines contain video modulation because some lines are blanked during vertical retrace.)

If we assume that the picture element is square and that the resolution, or definition, is the same both horizontally and vertically, we can calculate the maximum frequency produced. The ratio of the width of the picture to its height has been set at 4:3. This is known as the aspect ratio. If the vertical resolution is set by the 480 horizontal lines, the number of dots in one horizontal line for equal resolution horizontally would be 480 × 4/3. or 640. Multiplying the number of horizontal lines (480) by the number of dots in one horizontal line (640), we obtain a possible maximum of 307,200 picture elements for a single frame. Thirty frames per second (two 60-Hz fields) multiplied by this figure of 307,200 produces an upper frequency of 9,216,000 picture elements per second.

Fortunately, this figure does not represent the top frequency, as you can see from the video signal produced by scanning a checker board (Fig. 9-2) consisting of white and black squares, 480 squares vertically and 640 horizontally. In the television transmission system that is standard in the United States, an increase in carrier strength, caused by an increase in video-modulating voltage, produces a darker spot on the cathode-ray tube. The brightest spot in the picture corresponds to the lowest modulating voltage. Thus, when the scanning beam in the camera tube reaches the edge of a black square at point 1 in Fig. 9-2, the video-modulating voltage of the transmitter assumes its maximum positive value and remains at this value as the scanning action crosses the black portion of the image. At point 2, the scanning reaches a white area, and the modulating voltage changes to its maximum negative value.

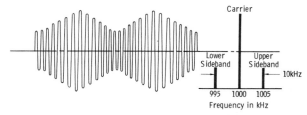

Fig. 9-1. Double-sideband modulated carrier wave; 50-percent modulation, 5000-Hz tone.

The video voltage then remains negative until point 3 is reached, at which point the cycle begins to repeat. Since two picture elements have been scanned to produce one cycle of video voltage, the top modulating frequency required will be one-half the number of picture elements, or, in this example, 4,608,000 Hz (4.6 MHz) for 9,216,000 picture elements. Since we used as an example the maximum resolution of the system, in which a picture element has been made the same size as the scanning spot, the video output would be the sine wave at C, rather than the square wave at B in Fig. 9-2.

If double-sideband modulation of the type employed in am radio broadcasting were used in tele-

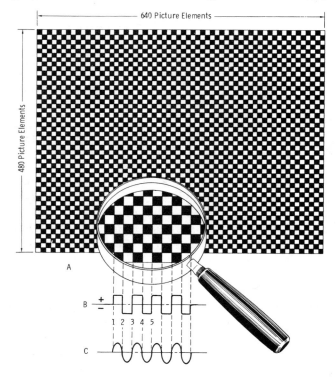

Fig. 9-2. Video modulation produced by scanning a checkerboard in which the picture elements are equal to the limit of the system resolution.

vision, sidebands extending 4.5 MHz on both sides of the carrier would require a bandwidth of 9.0 MHz for the video signal. In the early days of television research, when the pictures were of much lower definition (90 lines instead of 525), double-sideband modulation was employed. In addition to the bandwidth required for the video carrier, the accompanying sound carrier and a guard space must be provided. The Federal Communications Commission has assigned a channel space of only 6 MHz for each television channel. Therefore, it is impossible to use the double-sideband type of modulation and still retain the definition that is possible in a 525-line picture. Other types of modulation which require less channel space to accommodate the wide band of modulation frequencies encountered in television are (1) single-sideband modulation and (2) vestigial-sideband modulation (one sideband plus a vestige of the other).

In the process of detecting (demodulating) an amplitude-modulated carrier, one sideband is always eliminated, and the information in the other is amplified and used to produce the picture. The single-sideband method requires only half the space in the radio spectrum for a given maximum modulating frequency, compared with the double-sideband method. Single-sideband modulation is produced by passing a double-sideband modulated carrier through a radio-frequency filter network, which suppresses the undesired sideband. Although the single-sideband method allows the greatest use of channel space, it is impractical for transmission of television signals and has been replaced by the vestigial-sideband method because:

1. The type of filter which must be employed to completely suppress one sideband introduces serious phase distortion of the low-frequency components of the video signal and results in a blurred picture.
2. Even though means of correcting the phase shift may be developed, the single-sideband method would require extremely accurate tuning of the receiver and a high degree of freedom from oscillator-frequency drift. Otherwise, the low-frequency portion of the video modulation would be lost.

VESTIGIAL-SIDEBAND VIDEO MODULATION

To overcome the difficulties encountered with single-sideband modulation, the system of modu-

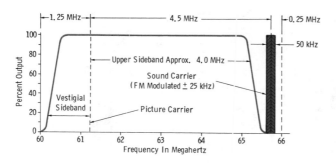

(A) Channel 3 output characteristic.

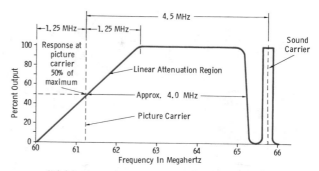

(B) Ideal receiver response characteristic.

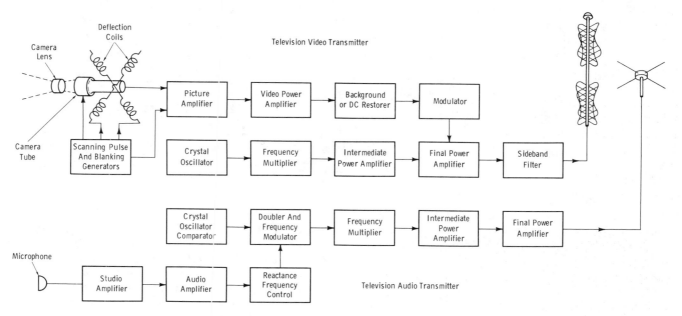

(C) Essential elements of a television transmitter.

Fig. 9-3. Output characteristic of a television transmitter, the ideal response characteristic, and a block diagram of the essential elements of a television transmitter.

lation employing all of one sideband and a vestige of the other has been adopted as standard in the United States. The fact that a part, or vestige, of one sideband is transmitted gives it the name vestigial-sideband modulation.

Fig. 9-3A shows a typical television channel employing vestigial-sideband modulation. Channel 3 has been chosen for this example. A block diagram of the essential parts of the transmitter required to produce this type of carrier and modulation is shown in Fig. 9-3C.

In this channel, which extends from 60 to 66 MHz, the picture-carrier frequency is placed 1.25 MHz above the lower limit of the channel (at 61.25 MHz). Video modulation of this carrier produces an upper sideband with frequencies ex-

tending to approximately 4 MHz. When color is being transmitted, the chroma signal is considered part of the video signal. The chroma subcarrier is located 3.58 MHz above the video carrier, and the chroma sidebands extend to slightly more than 4.0 MHz. The accompanying sound has a carrier frequency 4.5 MHz above the video carrier (at 65.75 MHz). A guard channel or space of 0.5 MHz separates the video modulation from the sound channel. The sound produces a frequency modulation of its carrier with a maximum deviation of 25 kHz (a total carrier swing of 50 kHz). Between the sound carrier and the upper end of the channel is another guard band of 0.25 MHz.

Note that the lower-sideband frequencies are extended to approximately 1 MHz below the video-

P = PICTURE CARRIER FREQ.(MHz)
S = SOUND CARRIER FREQ.(MHz)

Picture (P) / Sound (S) MHz	Channel	Freq. Limits (MHz)
P 55.25 / S 59.75	2	54–60
P 61.25 / S 65.75	3	60–66
P 67.25 / S 71.75	4	66–72
P 77.25 / S 81.75	5	76–82
P 83.25 / S 87.75	6	82–88
P 175.25 / S 179.75	7	174–180
P 181.25 / S 185.75	8	180–186
P 187.25 / S 191.75	9	186–192
P 193.25 / S 197.75	10	192–198
P 199.25 / S 203.75	11	198–204
P 205.25 / S 209.75	12	204–210
P 211.25 / S 215.75	13	210–216
P 471.25 / S 475.75	14	470–476
P 477.25 / S 481.75	15	476–482
P 483.25 / S 487.75	16	482–488
P 489.25 / S 493.75	17	488–494
P 495.25 / S 499.75	18	494–500
P 501.25 / S 505.75	19	500–506
P 507.25 / S 511.75	20	506–512
P 513.25 / S 517.75	21	512–518
P 519.25 / S 523.75	22	518–524
P 525.25 / S 529.75	23	524–530
P 531.25 / S 535.75	24	530–536
P 537.25 / S 541.75	25	536–542
P 543.25 / S 547.75	26	542–548
P 549.25 / S 553.75	27	548–554
P 555.25 / S 559.75	28	554–560
P 561.25 / S 565.75	29	560–566
P 567.25 / S 571.75	30	566–572
P 573.25 / S 577.75	31	572–578
P 579.25 / S 583.75	32	578–584
P 585.25 / S 589.75	33	584–590
P 591.25 / S 595.75	34	590–596
P 597.25 / S 601.75	35	596–602
P 603.25 / S 607.75	36	602–608
P 609.25 / S 613.75	37	608–614
P 615.25 / S 619.75	38	614–620
P 621.25 / S 625.75	39	620–626
P 627.25 / S 631.75	40	626–632
P 633.25 / S 637.75	41	632–638
P 639.25 / S 643.75	42	638–644
P 645.25 / S 649.75	43	644–650
P 651.25 / S 655.75	44	650–656
P 657.25 / S 661.75	45	656–662
P 663.25 / S 667.75	46	662–668
P 669.25 / S 673.75	47	668–674
P 675.25 / S 679.75	48	674–680
P 681.25 / S 685.75	49	680–686
P 687.25 / S 691.75	50	686–692
P 693.25 / S 697.75	51	692–698
P 699.25 / S 703.75	52	698–704
P 705.25 / S 709.75	53	704–710
P 711.25 / S 715.75	54	710–716
P 717.25 / S 721.75	55	716–722
P 723.25 / S 727.75	56	722–728
P 729.25 / S 733.75	57	728–734
P 735.25 / S 739.75	58	734–740
P 741.25 / S 745.75	59	740–746
P 747.25 / S 751.75	60	746–752
P 753.25 / S 757.75	61	752–758
P 759.25 / S 763.75	62	758–764
P 765.25 / S 769.75	63	764–770
P 771.25 / S 775.75	64	770–776
P 777.25 / S 781.75	65	776–782
P 783.25 / S 787.75	66	782–788
P 789.25 / S 793.75	67	788–794
P 795.25 / S 799.75	68	794–800
P 801.25 / S 805.75	69	800–806
P 807.25 / S 811.75	70	806–812
P 813.25 / S 817.75	71	812–818
P 819.25 / S 823.75	72	818–824
P 825.25 / S 829.75	73	824–830
P 831.25 / S 835.75	74	830–836
P 837.25 / S 841.75	75	836–842
P 843.25 / S 847.75	76	842–848
P 849.25 / S 853.75	77	848–854
P 855.25 / S 859.75	78	854–860
P 861.25 / S 865.75	79	860–866
P 867.25 / S 871.75	80	866–872
P 873.25 / S 877.75	81	872–878
P 879.25 / S 883.75	82	878–884
P 885.25 / S 889.75	83	884–890

Fig. 9-4. Channel allocations for commercial television.

carrier frequency. For video modulation frequencies extending to 1 MHz, the transmitting system is essentially double sideband. The presence of the lower sideband would overemphasize the low frequencies if the receiver response characteristic were flat throughout the entire band. To compensate for this overemphasis, the receiver characteristic is such that the carrier suffers a 50% loss (Fig. 9-3B). The receiver response is made to drop off linearly from a frequency 1.25 MHz above the video-carrier frequency to the lower limit of the band. When the receiver is correctly tuned to the channel, this loss of the lower sideband produces a flat output from approximately 25 Hz to 4.5 MHz.

When the Federal Communications Commission first announced channel allocations, 13 channels were assigned for television broadcasting. These channels were divided into a low-band and a high-band group, with each channel occupying a 6-MHz band. Channels 1 through 6 originally covered a band of 44 to 88 MHz (72 MHz to 76 MHz being reserved for other services). Channels 7 through 13 covered a band of 174 to 216 MHz. Channel 1 (44 to 50 MHz) is no longer used for television. In 1952, the uhf band was opened by the FCC. This band includes 70 channels (14 through 83) between 470 and 890 MHz. The vhf and uhf television channels, with their respective picture and sound carriers, are listed in Fig. 9-4.

The 13 higher uhf channels (71 through 83) are no longer used for television broadcasting. This portion of the frequency spectrum has been given over to the land-mobile communication services. However, in some remote areas, these uhf channels are used for low-power television translators. (A television translator is used to receive a signal on one channel and retransmit it on another channel to valleys and similar areas which cannot receive a direct channel.)

Fig. 9-5 illustrates the consecutive arrangement of channel spacing. Note the guard space of 0.25 MHz between the sound carrier of the adjacent television channel and the low end of the desired channel. Similarly, there is a space of 0.25 MHz

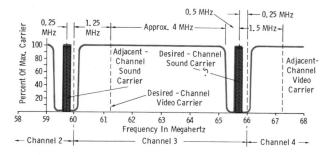

Fig. 9-5. Arrangement of consecutive channel spacing for Channels 2, 3, and 4.

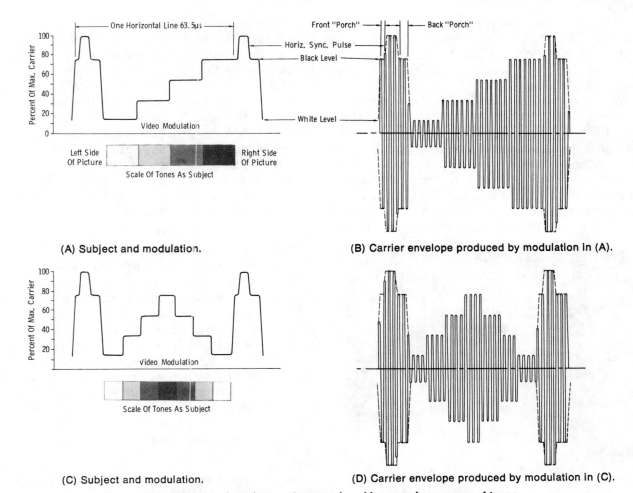

(A) Subject and modulation.

(B) Carrier envelope produced by modulation in (A).

(C) Subject and modulation.

(D) Carrier envelope produced by modulation in (C).

Fig. 9-6. Video signal and carrier envelope produced by scanning a range of tones.

between the sound carrier of the desired channel and the low end of the video modulation of the next higher channel. Television receivers must be able to attenuate or reject these undesired transmissions occurring in adjacent channels.

THE VIDEO SIGNAL

Fig. 9-6 shows the video signal which would be produced if a card carrying a series of tones ranging from pure white, through a number of grays, to black were placed in front of the television camera. Fig. 9-6A represents one horizontal line scanned across the series of tones. The line is preceded and followed by pedestals upon which the horizontal-synchronizing pulses are mounted. A scale at the left of Fig. 9-6A indicates the percentage maximum of the carrier corresponding to the various light values. The region from 75 to 100% is reserved for the sync pulses; therefore,

the 75% point is the blackest part of the picture. The region beyond this black point is known in television slang as blacker than black and in engineering circles as the infrablack region.

At the other end of the modulation scale (from 10 to 15% of maximum amplitude), the brightest highlight, or white region, is found. FCC regulations specify that this white level shall not exceed 15% of maximum carrier amplitude. The basic method of receiver design now employed, known as intercarrier sound, requires a white level of at least 10% of maximum carrier. For this reason, television broadcast stations must now hold the white level within quite narrow levels.

The intermediate gray tones fall between these two extremes (Figs. 9-6A and 9-6C). Fig. 9-6A shows a series of tones in which the white portion of the picture is at the left and the black portion is at the extreme right. Fig. 9-6C shows a succession of white through black to white again; the black level or pedestal is reached at the center of the scan. The television video-carrier envelopes that would be produced by the video-modulating waves of Figs. 9-6A and 9-6C are shown in Figs. 9-6B and 9-6D.

An actual television subject will be represented by continuously varying tone values along each of the 470 to 480 scanning lines. Fig. 9-7 illustrates a typical television image. In this figure, one of the horizontal lines (X-X) has been analyzed for change of video signal due to the major differences in reflected light intensity from the subject. The variation of camera signal due to scanning of line X-X causes the video-modulating voltage that

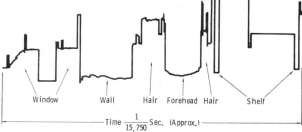

Fig. 9-7. Variation of video voltage produced by scanning one line (X-X) in the television scene.

is drawn at the bottom of Fig. 9-7. The sudden changes caused by crossing from bright areas to dark areas represent rapid changes of the video voltage. For faithful reproduction of such sudden changes, the system must have extremely wide-band response characteristics. Gradual variations, like those represented by the slight variations in the flesh tones of the face and by gradual transitions in the background, can be accommodated by middle frequencies of the video band.

A number of times in our discussion of video modulation, we have indicated that an increase of carrier strength produces a darker picture until the spot on the picture tube is finally extinguished and a block dot is produced. This is known as negative modulation. Negative modulation was chosen by the National Television Systems Committee (a radio industry group which acted as an advisory committee to the FCC) after a trial of positive and negative systems in the laboratory and in experimental field demonstrations. The reasons for the choice of negative transmission as the United States standard are:

1. Static and noise in the receiver and man-made noise from automobiles, oil-burner ignition, electric shavers, etc., produce black spots on the picture in negative transmission but would produce bright flashes of light if the video modulation were positive. The former condition is the least annoying.

2. In the negative system, both horizontal- and vertical-synchronizing pulses are at maximum carrier amplitude. Thus, receiver synchronization with the transmitter is ensured even during low signal strength at the limits of the reception area.

THE DIRECT-CURRENT COMPONENT OF THE VIDEO SIGNAL

Video modulation differs from the audio modulation of sound broadcasting because it is a varying direct current, rather than the familiar alternating current of speech and music modulation.

The direct-current component, or bias, corresponds to the average illumination of the scene being televised. This is equivalent to an average of the camera output for all the lines comprising a frame-scanning interval. The camera tube produces an alternating voltage output, which is proportional to the variation in brightness of the parts of the picture being scanned. Since the output of the camera tube (such as the vidicon) is capacitively coupled to the input grid of the camera amplifier, any direct-current components of the camera-tube output cannot be passed on to the succeeding stages. The remaining video-amplifying stages of the transmitter are also capacitively coupled and, therefore, cannot amplify the direct-current component.

Direct-current components in the video signal are due to those portions of the scene being televised which have no change in brilliance over part of the horizontal line. A uniform gray background, for instance, produces a uniform single value charge on that part of the mosaic representing its image. As the scanning beam of the camera tube crosses this section of the mosaic, the output voltage does not change; consequently, no alternating voltage is passed on to the camera amplifier input.

If the average lighting of the scene or background were not taken into account by adding the correct dc bias to the video modulator, the contrast between the various parts of the picture would be correct, but the background illumination or shading would not.

The following examples of typical subjects will illustrate the effect on the received image if the average or dc component were not transmitted.

1. A dancer dressed in a white costume and with a black curtain as a background is chosen as the subject. Assume that the television system, from camera tube to the receiving picture tube, has been so adjusted that the dancer's costume is rendered on the picture tube as a satisfactory highlight, and the black curtain appears as black. What will happen if several more dancers, similarly garbed in white, enter the scene? If the dc component representing the new average light value of the scene is in the transmitted carrier and is restored at the picture tube, the scene will be reproduced correctly. If the new dc bias were not provided at the transmitter, the highlights of the dancers would be gray, and any gray areas in the background would disappear into the black area beyond cutoff.

2. As an opposite extreme, consider a hockey arena as the subject. The system has been set up for proper rendition of the ice as the highlight value. If the opposing teams now skate into the field of view, this large area of dark figures will degrade the highlight tones of the ice, and the figures will not be as dark as the actual scene contrasts require.

To help understand the reasons for the effects described in the two examples, we will show two types of subject material which, in the absence of

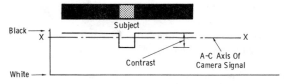

(A) Scanning a gray on a black background.

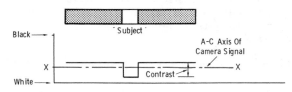

(B) Scanning a white on a gray background.

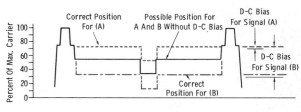

(C) Video modulation.

Fig. 9-8. Examples of why direct-current components of video signal are needed.

the dc component, would produce the same video signal and cause confusion in the reproduced image.

Fig. 9-8A shows the video signal from the camera amplifier which is caused by a gray line being scanned on a black background. Since no dc component is present, the alternating current is averaged about the line X-X. Fig. 9-8B shows the camera amplifier output when a white line across a gray background is scanned. The contrast between the line and the background is the same in each instance, and the ac video modulation is also the same. When the video signal in each example is referred to the black level (Fig. 9-8C), the light values are placed at their proper points on the scale, and there is no confusion. The direct-current component, caused by the average value of the background, now differentiates between the gray line on the black background and the white line on the gray background.

At the television transmitter, the dc component is often added to the video signal in the camera amplifiers, but it is subsequently lost when the signal passes through higher-level capacitively coupled video amplifiers. The dc component is then reinstated at various points in the video amplifiers through the use of clamping circuits.

QUESTIONS

1. What is the range of modulating frequencies encountered in the television video signal?

2. What is the aspect ratio of a television picture?

3. When a change in the video-modulating voltage causes a momentary increase in carrier strength at the transmitter, what happens to the brightness of the spot on the television screen at the receiver?

4. How many megahertz wide is each television channel?

5. Is double-sideband or vestigial-sideband modulation used in transmitting the video signal?

6. Where is the picture carrier placed in the television channel? What is the limit of the upper sideband of the video signal?

7. What is the maximum percentage of video modulation? What is the specified white-level limit?

8. Are the horizontal-sync pulses transmitted at maximum or minimum carrier amplitude?

CHAPTER 10

Sync-Pulse Separation, Amplification, and Use

Sync-pulse separation has been mentioned previously in connection with deflection systems. At this time we will consider in greater detail the separation, amplification, and use of sync pulses. We will begin our study with a discussion of sync-pulse separation.

SYNC-PULSE SEPARATION

The horizontal- and vertical-synchronizing pulses and their time relationship to the scanning sawtooth were described and illustrated in Chapter 7. These pulses occur when the electron beam in the picture tube is cut off. How the sync pulses are separated from the video signal and used to control the horizontal- and vertical-scanning systems of the receiver will be discussed. Specifically, we shall consider the separation of the horizontal pulses occurring at the end of each of the 525 lines in the picture. The vertical pulses will be covered when we consider the methods of segregating the vertical pulses from the horizontal pulses.

The pulses can be clipped from the signal at three places in the circuit:

1. At the video-detector input.
2. At any of the video-amplifying stages.
3. At any point of restoration of the average background light of the picture.

Fig. 10-1C shows the video signal with its picture and synchronizing information as it appears at the input of the video detector (Fig. 10-1A). This signal consists of horizontal pulses (as detailed in Fig. 7-1) mounted on a shelf or pedestal. These are identified in Fig. 10-1C as (1) the pulse

and (2) the pedestal. Between the edges of the pedestals, the video-modulated envelope (3) of the carrier is found. This envelope represents the variations in light and dark of the video signal that modulates the picture-tube beam. Fig. 10-1B shows the form of the detected or demodulated wave; the sync pulse and pedestal are at (5) and (4), respectively. The picture information, or video signal, is shown in the part of the wave at (6).

Note that the sync pulses are at the top of the signal. Since these pulses occur when the picture tube is black, the darker tones of the picture are just below the pedestal, or point (4) of Fig. 10-1B. The signal, after it passes through the video amplifier, must reach the grid of the picture tube in such a phase that the sync pulses are the most negative part of the wave. This action accomplishes blanking during the return trace. We have seen that the polarity of the sync pulse, as it arrives at the grid of the scanning oscillator, must be of the proper polarity to ensure control. For this reason, the method of sync-pulse separation selected in any particular receiver design will depend on (1) the number of stages of video amplification and (2) the point in the circuit at which sync separation is accomplished.

All the methods of separating the synchronizing pulse from the rest of the video signal involve the fact that, in transmission, the pedestal or blanking level is always maintained at a definite point on the carrier wave (75% of maximum carrier). Therefore, the sync pulses occupy the top 25% of the wave. The problem of sync clipping thus resolves itself into one of amplitude separation or

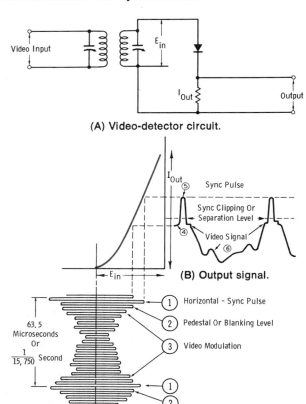

(A) Video-detector circuit.

(B) Output signal.

(C) Input signal.

Fig. 10-1. Demodulation (detection) of video carrier.

of removing the top 25% of the wave without passing the lower portion containing the video signal. The various methods that can be employed will be covered under the types of circuits used.

Diode Sync-Separation Circuits

Fig. 10-2 shows three diode sync-separation circuits. The input signal of each circuit is the composite video signal, and the output is a pulse signal that represents the sync pulses.

In Fig. 10-2A the video signal is coupled to the anode of the diode. The output signal is developed across a resistor in the cathode circuit. In order to keep the diode at cutoff until sync-pulse time, the anode is connected to a negative delay-bias source. When the highly positive sync pulses come along, the delay bias is overridden and the diode is allowed to conduct. During this time, a voltage is developed across the cathode resistor with the polarity as shown. The output taken across the cathode resistor is in the form of positive pulses.

Fig. 10-2B is an inverted version of the same circuit. In this circuit, the video signal is coupled to the cathode, and the output signal is developed

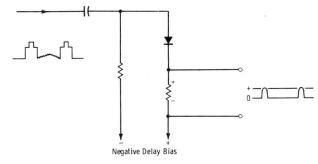

(A) Positive sync pulses with delay bias on diode.

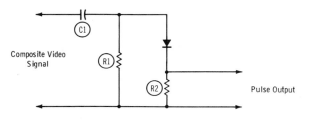

(B) Negative sync pulses with delay bias on diode.

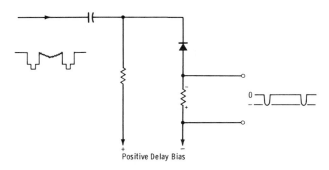

(C) Positive sync pulses with self-biased diode.

Fig. 10-2. Diode sync-separation circuits.

at the anode. Cutoff is accomplished by a positive delay bias applied to the cathode. The input signal is of opposite polarity to that applied to the diode in Fig. 10-2A. The sync pulses are highly negative. When the sync pulse comes along, the positive delay bias is overridden and the diode conducts. A voltage is thus developed across the anode resistor with the polarity shown. The output taken across the anode resistor is in the form of negative pulses.

The circuit in Fig. 10-2C is similar in operation to the one in Fig. 10-2A, except that the bias required for correct delay of the diode action is obtained from the charge placed upon C1 during diode conduction. A negative voltage is developed at the anode of the diode by the discharge of C1 through resistor R1. The time-constant used in bias circuit R1-C1 is long, compared to the horizontal and vertical scanning times. Thus, the diode can be biased automatically by the video signal to the proper point, so that the sync pulses are clipped from the signal.

In most modern tube-type receivers, triodes or pentodes are employed, rather than the diode circuits just discussed. The choice of tube type depends on the sync-pulse amplitude requirement, polarity requirement, and the point in the circuit at which separation is accomplished. Generally, these tubes allow some voltage gain to be realized in the separation process. In certain circuits, leveling or limiting can also be realized. Three basic types of triode sync-separation circuits are shown in Fig. 10-3.

The circuit of Fig. 10-3A uses grid rectification of the video signal to bias the control grid so that plate-current cutoff occurs at the desired pedestal level. This action is similar to the one just discussed for the diode circuit in Fig. 10-2C. Two additional actions are found in this circuit.

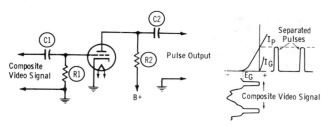

(A) Grid cutoff by signal rectification.

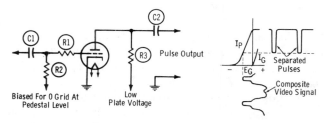

(B) Separation by positive grid and plate-current saturation.

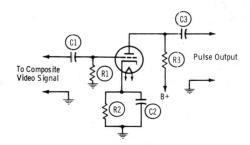

(C) Cathode-bias circuit.

Fig. 10-3. Triode sync-separation circuits.

1. The sync pulses are large enough to drive the grid positive, and the lowered grid resistance limits the input signal by loading.
2. Some amplification of the sync pulse occurs because of the amplifying properties of the triode.

In Fig. 10-3B, the operating conditions are quite different from those of the circuit in Fig. 10-3A. The tube is biased from an external voltage source, through resistors R1 and R2, until it just starts to draw grid current at the black or pedestal level. The input signal is inverted in polarity from that of the preceding example. The sync-pulse portion of the video-input signal is the most negative. The plate voltage is made so low that the plate current is saturated at a near zero grid voltage. The portion of the video-input signal which is more positive than the desired clipping level lies in the saturation region and produces no further rise in plate current. For this reason, limiting or leveling occurs at this saturation point. The negative grid voltage from the sync-pulse portion of the input signal causes a drop in plate current, as shown in the waveform drawings of Fig. 10-3B. To better illustrate circuit action, the amplitude of the sync input pulses has been limited in the drawing. If the amplitude were extended to beyond the grid cutoff point, limiting would also occur because of plate-current cutoff, and the output pulses would still be uniform but larger in size. Series resistor R1 in the grid circuit limits the amount of current that the grid can conduct over any one frame. This limiting prevents a long-time blocking condition from developing because of an excessive charge on capacitor C1.

The circuit in Fig. 10-3C employs cathode bias to establish the correct operating point for plate-circuit separation of the sync pulses. The values chosen for resistor R2 and capacitor C2 are determined by the following considerations:

1. The resistor must have such a value that the plate-current pulses above the clipping level will produce a voltage drop equal to the required operating grid bias. For high-mu tubes, this resistor will be around 10,000 ohms.
2. The value of capacitor C2 must be high enough to maintain constant bias voltage throughout at least one vertical blanking period, and yet low enough that it can change

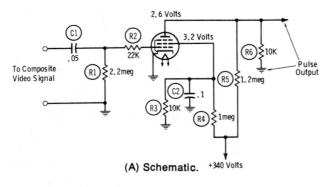

(A) Schematic.

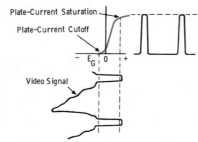

(B) Input and output waveforms.

Fig. 10-4. Pentode sync-separating circuit that incorporates leveling or limiting.

its charge as the average background lighting of the scene changes.

Pentode Sync-Separation Circuits

Fig. 10-4 illustrates the use of a sharply cutoff pentode as a combination sync separator and limiting amplifier. Plate-current saturation is ensured by operating both the screen and plate at extremely low voltages. This low voltage is accomplished with dividing networks. The screen network consists of resistors R3 and R4. With a B+ voltage of 340 volts, the screen is held at the extremely low voltage of 3.2 volts. The plate-supply network of resistors R5 and R6 maintains the plate at 2.6 volts. Plate-current saturation occurs just after the grid voltage goes positive, as shown in the diagram in Fig. 10-4B. The grid circuit network, resistors R2 and R1 with capacitor C1, establishes the operating point of the circuit by grid-circuit rectification of the video signal. This rectification ensures plate-current cutoff just below the pedestal or black region. You can see from the drawings that the synchronizing pulses are clipped at both ends, and limiting occurs. Because a pentode under these conditions exhibits very low voltage gain, a circuit of this type must either be operated at a high signal level

or be followed with sync amplifiers.

Series grid resistor R2 has a function similar to that of resistor R1 in the positive-grid triode of Fig. 10-3B. However, R2 in Fig. 10-4A has an additional function. In the absence of resistor R2, sharp noise pulses (such as those due to automobile ignition, with amplitudes possibly higher than that of the sync pulses) would cause an excessive bias voltage on C1 and result in blocking. Loss of synchronization would occur while the tube was blocked.

Cathode-Follower Used With Sync-Separation Circuits

Fig. 10-5 shows a circuit which employs a cathode-follower ahead of a sync separator. The cathode-follower tube (V1) operates on the linear portion of the curve. This is due to the balance of

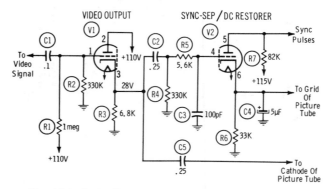

Fig. 10-5. Cathode-follower used with sync separator.

the bias across resistor R3 and the voltage from the B+ supply produced by network R1 and R2. A net negative voltage of −2 volts appears between grid and cathode.

Cathode-follower tube V1 functions both as video output and sync takeoff. The video output for the picture-tube cathode is taken directly from cathode load resistor R3 through capacitor C5. This same load resistor feeds a triode (sync separator and dc restorer) tube, V2, through the network of C2, R4, R5, and C3. The combination of R5 and C3 filters out the high-frequency components of the video signal.

Sync pulses are taken from the plate of triode V2, the operation of which is similar to the circuit in Fig. 10-3C. The dc bias voltage across network R6-C4 is used to restore the dc component of the picture signal responsible for the average light of the televised scene. This dc component has been lost in passing through the video-amplifier stages.

Gated Sync-Separation Circuits

The sync separators which have been described have one feature in common—the output signal is low in amplitude. Additional stages of amplification (and sometimes clipping) are needed to produce a sync signal of sufficient amplitude to trigger the sweep oscillators. A simpler type of separator circuit called the gated sync separator, has been developed. This circuit eliminates the extra amplifier stages because its tube is usually a pentagrid, which supplies many times more gain than a triode.

The circuit shown in Fig. 10-6 is built around one tube which does triple duty as a sync separator, agc amplifier, and noise inverter. The most unusual feature of this circuit is that the inverter removes noise pulses from the agc signal as well as from the sync signal. The agc amplifier circuit will be described in a later chapter.

A unique tube, the 6BU8, has been used in this circuit. The tube might be called a "Siamese twin" pentode. The cathode, the first control grid, and the screen grid are common to both sections of the tube. Outside the screen grid, there is a second control grid, which is split into two distinct sections; there are two separate plates in line with these two control grids. The currents in the two plate circuits can be controlled simultaneously by the first control grid, or separately by the pair of second control grids.

The cathode is grounded, so that variations in the plate current of one section of the tube will not be coupled to the other section. The sync-separator output is taken from the plate connected

to pin 3, and the agc output is obtained from the other plate. The first control grid (pin 7) is employed as a noise inverter.

A composite video signal containing positive-going sync pulses is coupled from the video-amplifier plate through C2 to one of the second control grids (pin 6) of the 6BU8 tube. Grid-leak bias is developed by C2 and R4. Plate current in the right-hand section of the 6BU8 is cut off, unless a sync pulse is on the grid. The waveform of plate voltage measured at pin 3 of the 6BU8 is typical of the output of a sync separator. Negative-going sync pulses are developed at this plate. They are applied to the horizontal-afc tube and vertical oscillator without further amplification. The vertical oscillator used with this circuit is a multivibrator and requires a negative sync pulse for proper triggering.

The first control grid of the 6BU8 tube receives a composite video signal from the video-detector output. This signal has the same waveform as the sync input signal on pin 6, but is opposite in polarity. One might assume that the two signals would cancel each other. Actually, the signal on the first grid has such a small amplitude that it has only a slight degenerative effect on the output of the tube. The action of the first grid becomes important only when noise is in the input signals.

The first control grid is biased so that the tips of the negative-going sync pulses will almost drive the grid into cutoff. Noise pulses in the composite video signal sometimes have a greater amplitude than the sync pulses. When one of these bursts of noise reaches the first control grid, the grid voltage is driven below cutoff, and all conduction within the 6BU8 tube ceases momentarily. At the same instant, a positive noise pulse appears at the second control grid, but the temporary interruption of plate current prevents a corresponding pulse from developing at the plate.

If the noise inverter were not in the circuit, strong noise pulses would get into the output signal of the sync separator. They would tend to cause random triggering of the sweep oscillators, and unstable synchronization would result.

Transistor Sync-Separator Circuits

A simple transistor sync-separator circuit is shown in Fig. 10-7. Due to the signal-developed bias across R1, the transistor conducts only on the portion of the composite signal above the blank-

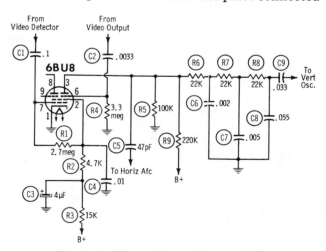

Fig. 10-6. The 6BU8 gated sync-separator circuit.

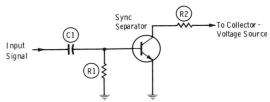

Fig. 10-7. Typical transistor sync-separator circuit.

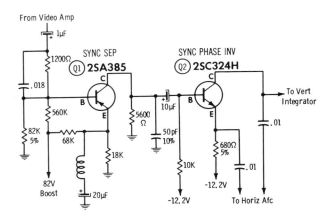

Fig. 10-8. Typical solid-state sync circuit using sync separator and phase inverter.

ing level. Therefore, only the sync pulses appear in the collector circuit of the transistor. The circuit action is similar to the diode circuit shown in Fig. 10-2C, except that the transistor provides amplification.

By limiting the value of the collector voltage, the transistor can be driven into collector saturation by strong incoming signals. In this manner, high-amplitude noise pulses are clipped in the collector circuit and prevented from reaching the sweep oscillators. Many transistor television receivers employ two or more sync-separator stages in order to provide proper clipping with strong signals.

SYNC-PULSE AMPLIFICATION, CLIPPING, AND SHAPING

Many television receivers employ more than one circuit to separate the sync pulses from the video signal. Additional stages are introduced to (1) invert the phase of the pulses (when not of the proper polarity to control the scanning oscillator); (2) clip the pulse width (for more reliable scanning control); (3) amplify the pulse (if it is not strong enough for control); and (4) level the pulse (to take care of variations in the video signal and minimize the effect of interfering noise pulses).

There has been little standardization in naming those stages, and we find the following descriptive titles in the service literature of various manufacturers: "sync clipper," "sync amplifier," "sync inverter," "sync leveler," "sync limiter," "pulse limiter," and "clamper." The various actions are self-explanatory; but it should be noted that, even though a sync stage is labeled as a sync amplifier, it is usually so biased that either cutoff or saturation contributes leveling or clipping as well as the desired voltage amplification.

Most modern tube-type receivers employ a circuit similar to Fig. 10-6 to perform sync-separating functions. In solid-state television receivers, the transistor sync-separator circuit is usually fol-

lowed by either a sync amplifier or a sync phase-inverter stage. In some cases, the phase inverter also serves as an amplifier, or vice versa, depending on which circuit configuration is employed. Fig. 10-8 shows a typical arrangement of sync separator and phase inverter. Here, the sync section is designed to supply a push-pull output to a balanced horizontal-afc circuit. In order to do this the last stage in the sync section is arranged as a phase inverter. Positive-going sync pulses are developed at the emitter, and negative-going pulses are developed at the collector. In other words, transistor Q2 operates as a phase splitter for the horizontal sync pulses. The circuit itself is a combination common-emitter and emitter-follower stage.

Another arrangement involves the use of sync amplifiers (or combination amplifiers and inverters) ahead of the sync separator, rather than following. A circuit such as this is shown in Fig. 10-9.

In many late-model television receivers the sync-separator function is incorporated in an integrated circuit (IC) which performs additional functions. In the circuit shown in Fig. 10-10, inte-

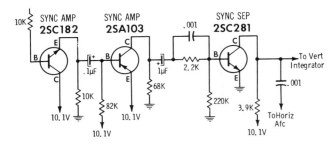

Fig. 10-9. Sync system providing amplification prior to the sync-separator stage.

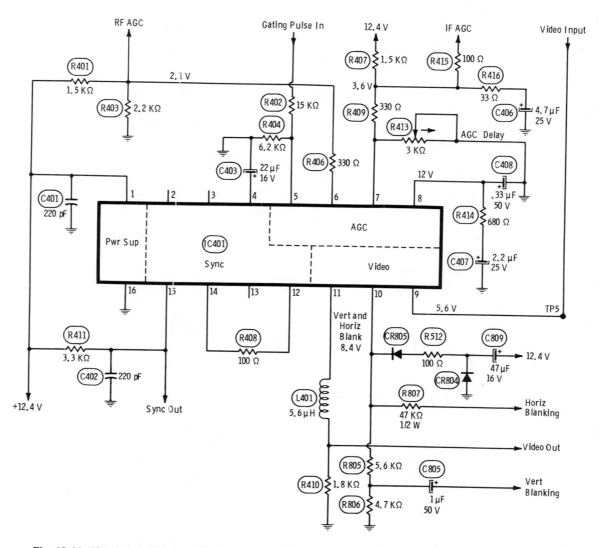

Fig. 10-10. Circuit using a single IC for agc, noise-gate, sync-separator, and sync-amplifier functions.

grated-circuit IC401 functions as a noise gate, develops the agc voltages, separates the sync information from the composite video signal, and amplifies the sync pulses. For our discussion in this chapter, we will be concerned primarily with the separation of the sync signal and the amplification of the horizontal and vertical sync pulses.

The composite video signal is fed into IC401 at pin 9. It is at this point that the IC performs its noise-gate function. When noise pulses that exceed the level of the sync tips appear at pin 9, the sync-separator function is cut off. This prevents the noise pulses from tripping the horizontal and vertical oscillators.

The integrated circuit, IC401, separates the video information from the sync information in-

ternally and only the sync pulses are amplified. Capacitor C403 is the sync-separator charging capacitor. It sets up the biasing that determines the clipping level for the sync pulses. The positive sync signal at pin 12 is coupled through R408 to pin 14 of the IC where it is inverted and amplified. A negative sync signal with a peak-to-peak amplitude of approximately 12 volts is taken from pin 15 of the IC and fed to the horizontal and vertical oscillators.

SORTING OF THE INDIVIDUAL HORIZONTAL AND VERTICAL PULSES

In the foregoing description of the various methods of separating the synchronizing pulses from the composite video signal, only the narrow horizontal pulses were mentioned. The longer vertical pulses are separated from the signal in the same process.

After the sync pulses have been removed from the video signal, the vertical pulses must be sorted from the horizontal pulses, and each must be fed to its respective deflection-scanning system. Since the horizontal and vertical pulses are equal in amplitude, the methods of separating them from the video signal cannot be used to distinguish between them. They do, however, differ in time duration. It is on this basis that sorting is accomplished.

Previously, we have mentioned differentiating networks for passing horizontal pulses and integrating networks for obtaining the vertical pulse. We will now consider the action of these systems in greater detail.

Horizontal-Pulse Separation

The horizontal pulses of the transmitted signal are approximately five microseconds in duration, as shown in Fig. 7-1. These pulses are impressed on a circuit of the type shown in Fig. 10-11C, which is known as an RC differentiating circuit.

Differentiation means the breaking down of a quantity into a number of small parts. The pulses in Fig. 10-11A are made into smaller parts, as shown in Fig. 10-11B, by the action of the circuit in Fig. 10-11C. The circuit consists of a capacitive and resistive combination in which the capacitor

is in series with the separated sync-pulse input and the resistor is shunted across the output. The time constant of this circuit is made short, compared with the duration of a horizontal-sync pulse. The sync pulse is held between 4 and 5 microseconds, and the time constant of the horizontal differentiating circuit is made between 1 and 2 microseconds. As described and illustrated in Fig. 5-7 (an RC circuit in which the time constant is short, compared with the duration of the applied square-wave pulse), the capacitor is completely discharged. A sharp pip of voltage occurs across the resistor at both the leading and trailing edges of the applied square-wave pulse.

The amplitude of the pip is determined not only by the amplitude of the square wave, but also by the steepness of the edge of the square wave. For this reason, the FCC limits the allowable slope of the leading and trailing edges. These slopes must not occupy more than 0.4% of the horizontal-line scanning interval of 63.5 microseconds.

The voltage pip due to the leading edge of the horizontal-sync pulse is shown as a positive pip at (1) in Fig. 10-11B. The dip due to the trailing edge of the horizontal pulse is shown as a negative voltage at (2). The leading-edge pulses are the ones which control the horizontal-scanning oscillator. The negative pulses are rejected by cutoff or saturation of one or more stages in the sync system.

When the longer-duration vertical synchronizing pulses arrive, the differentiating circuit acts as shown in Fig. 10-12. Here again, a positive pip occurs at the leading edge of each vertical pulse, and a negative pip occurs at the trailing edge. The leading-edge pulses continue to control the horizontal oscillator during vertical retrace. In this instance, however, two pulses occur during a

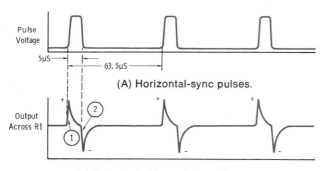

(A) Horizontal-sync pulses.

(B) Output of differentiating circuit.

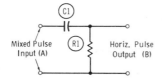

(C) RC circuit for horizontal separation.

Fig. 10-11. Horizontal-pulse separation, or differentiation.

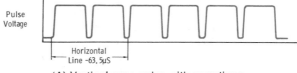

(A) Vertical sync pulse with serrations.

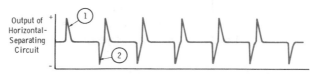

(B) Output of horizontal-differentiating circuit.

Fig. 10-12. Action of horizontal differentiating circuit on vertical-sync signal.

horizontal line-scanning interval. Only the first of these pulses is used to control the horizontal oscillator. The second pulse cannot cause lock-in, since it occurs while the oscillator is insensitive to tripping.

Vertical-Pulse Separation

In the description of vertical-scanning systems, we mentioned integrating networks for segregating the long-time vertical-field pulses from the sharp horizontal-line pulses. We will now consider the means of sorting these vertical-field scanning pulses from the composite scanning pulses and of using them to control the vertical-oscillator timing.

The integrating action which sorts the vertical pulses from the complex video signal is exactly opposite from the differentiation process for separating the horizontal pulses. Integration means the addition of a number of small elements to form a whole. Fig. 10-13C shows an integrating circuit. It is the opposite of the differentiation circuit in Fig. 10-11. The resistor is in series with the input, and the capacitor is connected across the output. The time constant of the combination

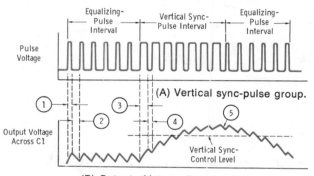

(B) Output of integrating circuit.

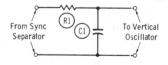

(C) Integrating circuit.

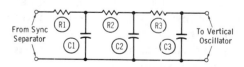

(D) Cascade integrating circuit.

Fig. 10-13. Vertical-pulse separation by integration.

is much longer than that employed for sorting the horizontal pulses. This time constant is made approximately equal to the duration of a horizontal pulse. Consequently, the charge accumulated by the capacitor, because of a horizontal pulse, is small and will decay rapidly. This action is shown in Fig. 10-13B. During the time shown at (1), the equalizing pulses produce only a small voltage across the capacitor. This voltage decays to near zero in the interval between pulses, as shown at (2). The much longer vertical-synchronizing pulses produce a greater charge in the capacitor during period (3). This charges does not completely decay during the short serration interval (4). Consequently, each vertical pulse adds an element of charge to the capacitor, and the voltage continues to build up during the interval of vertical pulses. The dotted horizontal line in Fig. 10-13B indicates the level at which this voltage becomes large enough to trigger the vertical-scanning oscillator. This point usually occurs after two or three vertical pulses have charged the capacitor.

The vertical-integrating network may be the two-element type shown in Fig. 10-13C, or a cascade network, as shown in Fig. 10-13D. The resultant time constant of the cascade network is smaller than that of any of the individual branches (R1-C1, R2-C2, or R3-C3). The overall time constant calculation is the same as for resistors in parallel. For the three-branch circuit in Fig. 10-13D with T1 for the time constant R1 × C1, T2 for R2 × C2, and T3 for R3 × C3, the effective circuit time constant (T) will be

$$\frac{1}{T} = \frac{1}{T1} + \frac{1}{T2} + \frac{1}{T3}$$

Individual time constants for a three-branch circuit are 30 to 60 microseconds. The effective overall circuit time constants are, therefore, between 10 and 20 microseconds.

The reasons for using cascaded integrating circuits are:

1. To prevent erratic control of vertical retrace by random noise or static pulses. Before such pulses could control the vertical oscillator, they would have to be comparable in duration and spacing to the vertical-sync pulses.

2. To smooth out the contour of the rising voltage wave (shown in the interval 3 to 5 of Fig. 10-13B) across the output capacitor. The action is similar to that of the familiar

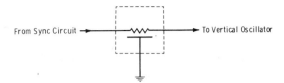

Fig. 10-14. Schematic representation of PC-type vertical-integrating network.

resistance-capacitance, power-supply filter system in which the ripple is reduced by successive stages.

Because of this smoothing action, an individual horizontal pulse cannot cause pairing of lines during retrace. The sections of the cascade network are not usually made with equal time constants. This unbalance prevents accidental triggering by noise pulses.

The vertical-integrating network used in modern television receivers usually has no more than two cascaded branches and is often the two-element branch shown in Fig. 10-13C. In many receivers, the vertical integrating network is a single, packaged component which includes the necessary capacitances and resistances. These PC networks are usually represented on the schematic as shown in Fig. 10-14.

THE FUNCTION OF VERTICAL-EQUALIZING PULSES

In Chapter 3 we briefly discussed interlaced scanning, which prevents flicker of the image. For simplicity, the retrace from bottom to top of the picture was shown as a straight line or single jump. Actually, the horizontal oscillator must be kept in step with the transmitter during vertical retrace, which lasts from 1250 to 1400 microseconds (20 to 22 horizontal lines). Fig. 10-15A shows a simplified version of the downward scanning, in which nine and one-half lines have been drawn to represent each field. Actually, a field consists of 262½ lines, less the lines lost during retrace. The first field, which starts at the upper left-hand corner (Point 1) and ends at the bottom center of the picture (Point 2), is shown by red lines. The second, or interlaced field, which starts at the top center (Point 3) and ends at the lower left-hand corner (Point 4), is shown by black lines. During vertical retrace, when the picture is blanked out, the beam moves upward under the combined action of both the vertical- and the hori-

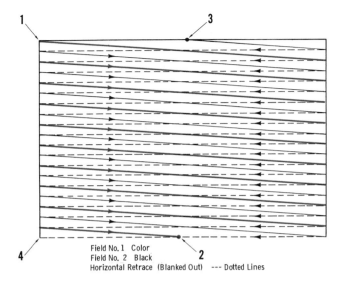

Field No. 1 Color
Field No. 2 Black
Horizontal Retrace (Blanked Out) --- Dotted Lines

(A) Active downward scanning.

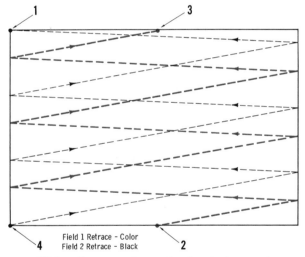

Field 1 Retrace - Color
Field 2 Retrace - Black

(B) Inactive upward scanning (vertical retrace).

Fig. 10-15. Simplified illustration of interlaced scanning.

zontal-deflection systems. This is represented in simplified form by the diagram in Fig. 10-15B. Here, three lines represent the twenty to twenty-two lines actually required during vertical retrace. Again, a red, dashed line represents the retrace of field No. 1, and a black, dashed line represents the retrace of field No. 2.

The dual functions of producing vertical retrace at the proper instant and of keeping the horizontal oscillator in synchronism are controlled by the equalizing and vertical pulses shown in Fig. 10-16. The vertical sync signal for the retrace of field No. 1 differs from that of field No. 2 in the spacing between the last horizontal pulse and the first

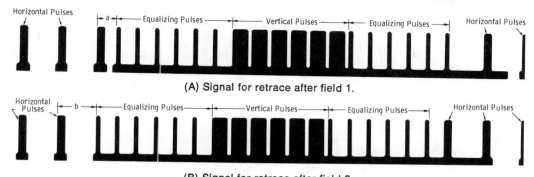

(A) Signal for retrace after field 1.

(B) Signal for retrace after field 2.

Fig. 10-16. Vertical-sync signals for retrace on alternate fields.

equalizing pulse. In Fig. 10-16A for field No. 1, this space (a) consists of only one-half of a horizontal line, since field No. 1 ends at the middle of the last line, as shown at Point 2 of Fig. 10-15A. In Fig. 10-16B for field No. 2, the space (b) between the last horizontal pulse and the first equalizing pulse consists of an entire horizontal line. Vertical blanking starts at the leading edge of the equalizing pulses. Thus, the successive field-blanking time is set up by the signal.

Even though retrace blanking is accurately established, vertical retrace may not take place at the proper instant unless the critical charge on the

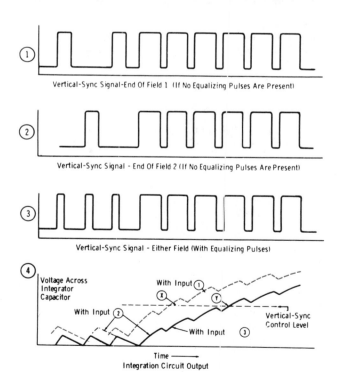

Fig. 10-17. Action of vertical-integrating circuit for alternate fields (with and without equalizing pulses).

integrating capacitor occurs at exactly the same point for each successive vertical-sync signal. How the equalizing pulses ensure this condition is shown in Fig. 10-17. At (1) the composition of a vertical-sync signal which would follow field No. 1, if the equalizing pulses were not present, is shown. This signal input to the integrating circuit would charge the capacitor, as shown by dotted line (1) on the charge curves of Fig. 10-17. This curve crosses the sync-control level at time X.

The vertical signal, without equalizing pulses for retrace at the end of field No. 2, would be as shown at (2) in Fig. 10-17. On the charge curves, the critical sync-control level would be reached at time Y, which is so much later than time X that proper interlace would not occur. When equalizing pulses are employed as shown at (3), the critical firing point for the vertical oscillator is at time Y for both fields. Successive fields, preceded by equalizing pulses, will therefore accurately control the oscillator and ensure proper interlace.

ACTION OF THE HORIZONTAL-DIFFERENTIATING CIRCUIT DURING THE VERTICAL PULSE

The formation of positive and negative pips at the leading and trailing edges, respectively, of the vertical sync pulses was described briefly. We will now consider, in detail, the action of the horizontal-differentiating circuit during the entire vertical pulse. Fig. 10-18A shows the pattern of the vertical signal following field No. 2. The output of the horizontal-differentiating circuit is shown in Fig. 10-18B.

The horizontal pulse which starts retrace of the bottom line of the picture is shown at (1) in Fig. 10-18A. The positive output pip produced by its

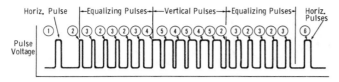

(A) Vertical pulse group following field 2.

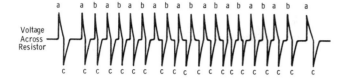

(B) Output of horizontal-differentiating circuit.

Fig. 10-18. Action of horizontal-differentiating circuits during vertical pulse period.

leading edge is shown at (a) in Fig. 10-18B. The pips (c) produced by the trailing edge of this horizontal pulse and of all the other pulses of the group are rejected by the sync system, as previously explained.

Each equalizing pulse (2 and 3) before and after the vertical pulses also produces a pair of positive and negative pips. Only the pips marked (a) are used for oscillator control. Those labeled (b) are rejected, since they occur during the scanning cycle when the horizontal oscillator is not sensitive to pulse control.

Each pulse of the vertical group (4 and 5) also produces a pair of positive and negative pips. However, only the positive pips (a) of Fig. 10-18B, are used. The horizontal pulse shown at (6) is one of a group occurring during the blanking period. The pips produced by the pulse at (6) are the same as those produced by horizontal pulse (1).

It is evident that the vertical-pulse group, because of the individual pulses and their different lengths, can ensure vertical retrace at the proper time and also keep the horizontal oscillator in step with the scanning in the camera tube at the transmitter.

QUESTIONS

1. In what three places in the television circuit can the synchronizing pulses be removed from the composite signal?

2. What is the function of a sync separator?

3. In a diode sync-separation circuit, what is the polarity of the output pulses when the input signal is applied to the anode of the diode?

4. What is the advantage of using a gated sync-separator circuit?

5. In solid-state receivers, what circuit normally follows the sync separator when the horizontal-afc circuit requires a push-pull input signal?

6. In a differentiator circuit, are the output pulses formed across a resistor or across a capacitor? Where are they formed in an integrator circuit?

7. What factor about the horizontal- and vertical-sync pulses makes it possible to distinguish between the two?

8. For a three-branch integrating circuit, is the total time constant smaller or larger than the individual time constant? How is the total time constant calculated?

EXERCISES

1. Draw a basic triode sync-separator circuit.

2. Show an integrator and a differentiator circuit, and draw the output waveform of each during the vertical sync-pulse period.

CHAPTER 11

The Receiving Antenna

Television receiving antennas are more critical in performance and play a more important role in the production of satisfactory reception than the antennas employed for am or fm broadcast reception. Some of the factors that must be considered in television reception are:

1. The nature of radiation and propagation of radio waves in the television bands.
2. Horizontal versus vertical polarization of television waves.
3. The wideband nature of the television channel and its relation to the susceptibility of the system to noise.
4. Ghost images due to multiple signal paths, caused by the reflection of the signal from buildings, mountains, hills, or other obstructions.
5. Ghost images due to reflections in the transmission line (lead-in), produced by a mismatch between the impedance of the line and that of the antenna or receiver input circuit.
6. The necessity for special types of antennas to achieve a desired directional reception pattern.
7. Wideband antennas to allow reception of stations widely separated in the television frequency spectrum.
8. Problems peculiar to fringe reception (reception of stations beyond the limits of the primary service area).

In our study of the television channel with its vestigial amplitude-modulated video and frequency-modulated audio components, we have seen that a channel width of 6 MHz is required for high-definition television. The modulation process requires that the carrier frequency be made at least ten times the highest modulating frequency. For this reason, the carrier frequencies of the television transmitters have been assigned in the portion of the radio spectrum above 50 MHz. The standard classification of the part of the radio spectrum between 30 and 300 MHz is vhf (very high frequencies). The propagation characteristics of vhf waves are considerably different from those of the lower frequencies employed for commercial am radio broadcasting. A review of how waves travel from the transmitter to the receiver will be helpful to the reader in understanding some of the transmission phenomena which occur in television bands.

The energy radiating from the transmitting antenna consists of two components—an electrostatic field and a magnetic field. These two fields are made up of lines of force at right angles to each other. The energy in the wave is divided equally between these two traveling alternating fields. Fig. 11-1 shows part of the wavefront of an electromagnetic wave in space. Since the wave is traveling in all directions from the transmitting antenna, the lines of Fig. 11-1 represent a small portion of a spherical surface.

POLARIZATION OF
THE TRANSMITTED WAVE

The direction of the lines of force of the electrostatic component defines the direction of polarization of the wave. In the graph of Fig. 11-1, the black horizontal lines are the electrostatic lines of force. Such a wave is said to be horizontally polarized with respect to the earth's surface. If the electrostatic lines are perpendicular to the earth's surface, the wave is said to be vertically polarized.

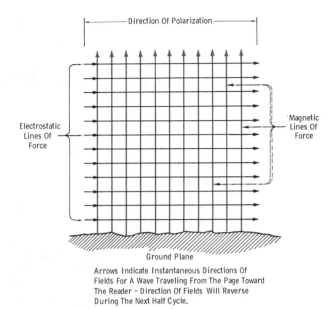

Fig. 11-1. Graphic representation of a horizontally polarized electromagnetic wavefront.

In the vhf portion of the television spectrum, the plane of polarization of the transmitted wave is the same as the position of the transmitting antenna with respect to the earth's surface; that is, a vertical antenna produces vertically polarized waves and a horizontal antenna produces horizontally polarized waves. The United States has adopted horizontal polarization as the standard for television transmissions. The reasons for choosing horizontal polarization are:

1. Most types of man-made interference, as well as interference from other radio communication transmitters, are vertically polarized; horizontal polarization helps reduce interference from these sources.
2. Horizontally polarized waves suffer less loss when reflected from the earth or when passing through the atmosphere.

In order for maximum power to be transferred between the transmitting and receiving antennas, both must be of the same polarization.

Although horizontal polarization has been the standard for television transmission in the United States for many years, a combination of horizontal and vertical polarization has recently been authorized by the FCC for commercial television stations. This combination of horizontal and vertical polarization is known as circular polarization. There are two distinct advantages to the employ-

ment of circular polarization. The main advantage is that ghost images due to reflected signals are substantially reduced. The second advantage is that the performance of portable television receivers with built-in vertical antennas is improved due to the vertical component of the circular polarized signal.

TYPES OF WAVE PATHS BETWEEN THE TRANSMITTER AND RECEIVER

Waves of different carrier frequencies follow different paths between the transmitter and the receiver. The waves are basically the same and would act alike in free space (beyond the earth's atmosphere). The earth's surface, the earth's atmosphere, and the presence of objects comparable in size to the length of the wave modify the transmission path as the frequency of the wave is changed.

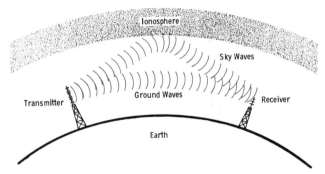

Fig. 11-2. The basic difference between ground-wave and sky-wave propagation.

Based on the paths they follow, radio waves can be classified into two basic types, ground waves and sky waves (see Fig. 11-2). The ground wave, which is of greater interest to us, includes all the components of a radio wave except those affected by the ionosphere or troposphere. This includes what is known as direct or space waves (line-of-sight) and ground-reflected waves (see Fig. 11-3). Waves reflected back to earth from the ionosphere or troposphere are considered sky waves.

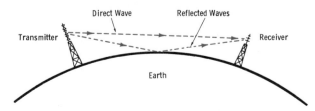

Fig. 11-3. Direct and ground-reflected waves.

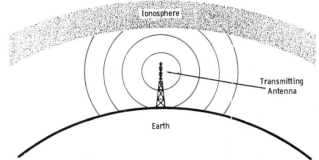

Fig. 11-4. Energy leaves the transmitting antenna in the form of wavefronts or shells of energy.

Ground Wave

The ground wave (or surface wave) moves along the surface of the earth and actually depends on the earth for a portion of its transmitting medium. The sky wave is radiated upward and is affected, more or less, depending on the frequency, by the ionosphere (ionized regions of the upper atmosphere). Actually, when we speak of ground waves, sky waves, etc., we are referring to portions of the same signal radiating from a particular antenna. The signal leaves the transmitting antenna in the form of spherical shells of energy, or wavefronts, as illustrated in Fig. 11-4. Certain portions of the signal move along the ground and can be detected by a receiving antenna, while other portions move outward into space and are considered wasted energy, except for refraction effects on certain frequency bands.

Since the ground wave, or certain elements of it, is the primary means of communicating with radio waves (especially at television frequencies), transmitting antennas are designed to extend, or increase, ground-wave coverage. The added coverage is referred to as antenna gain.

As stated previously, the primary portion of the ground wave depends on the earth's surface for its existence. The earth, however, offers resistance to the currents induced by the radiated signals, and this factor limits the distances that can be covered by primary ground-wave propagation. Ground-wave attenuation increases with frequency until, at approximately 30 to 40 MHz, coverage is only reliable in the immediate vicinity of the transmitter. Beyond these frequencies, the primary means of communication is with those portions of the ground wave which do not depend on the earth's surface for their existence. These are the direct line-of-sight or space waves and the ground-reflected waves illustrated in Fig. 11-3.

The action of frequencies in the range above approximately 40 MHz is often termed quasioptical, since the waves act in a manner similar to light rays. (Quasi in Latin means "as if" and optical means "light.") A transmitter operating in this frequency range, such as a television station, employs as high an antenna as is practical and economical. The carrier waves leaving the antenna act like the rays of light which would be produced if a powerful electric light were mounted on top of the antenna mast. The earth's curvature would cut off these light waves at the horizon. The horizon distance can be extended by erecting a tower for the observer of the light.

Waves emitted by television transmitters exhibit this line-of-sight action, as shown in Fig. 11-5A; however, the cutoff at the horizon is not sharply defined. In traveling through the air, the wave is bent slightly toward the earth, and the vhf range is considered to extend some 15% beyond the line-of-sight or optical horizon. Beyond this range, the strength of the signal decreases very rapidly.

Fig. 11-5B shows, by means of a nomograph, the effect of the height of the transmitting and receiving antennas on the optical horizon (shown at X) and on the vhf range (shown at Y). This chart is calculated for a smooth, spherical earth. If the transmitting and receiving antennas are at different heights above sea level, this difference should be taken into account when the vhf range is calculated. Large intervening objects such as buildings or hills can seriously reduce the signal level and thus the range of satisfactory reception. Fig. 11-5B indicates the advisability of locating the receiving antenna as high as possible.

Sky Wave

The sky wave, which accounts for long-distance reception on the low frequencies such as the broadcast and short-wave bands, is a wave that is bent back to the earth by ionized layers (ionosphere) in the upper atmosphere. Fig. 11-6 shows the effect of the ionosphere on transmitted waves. Some waves penetrate the ionized layer and others are refracted to the earth. Whether the waves are refracted or not depends on a number of conditions—the frequency of the wave, the density of the ionized layer, and the angle at which the transmitted wave enters the ionosphere.

The waves in Fig. 11-6 are all of the same frequency, and the ionosphere presents the same den-

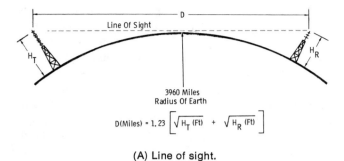

$$D(\text{Miles}) = 1.23 \left[\sqrt{H_T \text{(Ft)}} + \sqrt{H_R \text{(Ft)}} \right]$$

(A) Line of sight.

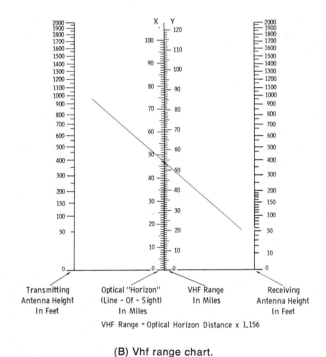

VHF Range = Optical Horizon Distance x 1.156

(B) Vhf range chart.

Fig. 11-5. Line-of-sight distance and vhf range as related to the height of the transmitting and receiving antennas.

sity for each wave. Because of their angle of entry, waves A and B in Fig. 11-6 can penetrate the ionosphere; therefore, they pass into space and are lost. The smaller the entry angle, the easier the wave is bent. A wave entering the layer at the angle of wave C cannot penetrate the ionosphere and is refracted back to the earth. The wave is steadily bent as it passes into the ionized layer and emerges toward the earth as though it had been reflected. The amount of bending that a wave will encounter is a function of the frequency or wavelength. At each frequency there is a critical angle at which the required density for refraction of the wave is obtained (Fig. 11-6). All waves steeper than wave C will pass through

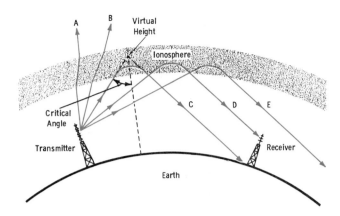

Fig. 11-6. The bending of radio waves back to earth. (Occurs at frequencies below approximately 40 MHz.)

the ionosphere. Those with more gradual slopes (to the right) will be reflected to the earth.

Frequencies of more than 30 to 40 MHz are not returned by the ionosphere, except under unusual sporadic conditions, because the higher frequency waves will easily penetrate the ionized layer. Therefore, this type of wavepath cannot be depended on for television transmission. Sometimes, however, abnormal atmospheric conditions, or disturbances in the earth's magnetic field, cause the density of one or more ionic layers to increase to such an extent that even these higher frequency waves are returned to earth, providing television reception from distant stations. Such a condition often produces what is known as co-channel interference, where two different stations in different areas are received simultaneously on the same channel. Interference of this type usually occurs in fringe areas where the signal from the primary station is weak.

WAVELENGTHS OF THE TELEVISION CHANNELS

In the ranges of frequencies assigned to television transmission (54 to 88 MHz, 174 to 216 MHz, and 470 to 890 MHz), the length of the electromagnetic wave is short when compared with the height of obstructions, such as buildings. This accounts for some of the peculiar transmission phenomena encountered. These will be described when we examine the effect of reflections.

The television service technician should become familiar with the actual wavelengths (in feet) of the television carrier frequencies in his locality, since the lengths of antenna elements are directly

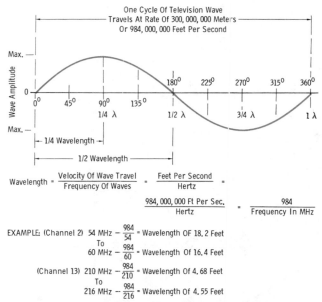

Fig. 11-7. Relationship of carrier frequency to wavelength of television signal.

related to the wavelength. For this reason, a review is needed of some of the relationships between the frequency of an electromagnetic wave and its wavelength.

All electromagnetic waves (cosmic, X rays, ultraviolet, light, infrared, and radio waves) travel through space at the same speed—approximately 300,000,000 meters, or 186,000 miles, per second.

The distance an electromagnetic wave travels in one cycle (360° of its sine-wave oscillation) is called one wavelength. Fig. 11-7 shows the relationship of wavelength, velocity, and frequency. When the velocity is in feet per second and the carrier frequency is in hertz, the numerator of the fraction in Fig. 11-7 can be divided by one million. The simple expression, 984 divided by the frequency in megahertz, is the result. The wavelengths encountered in the vhf television bands are from 4.6 feet (1.4 meters) to 18.2 feet (5.5 meters).

A handy rule-of-thumb from Fig. 11-7 is that the television wave travels approximately 984 feet (300 meters) in one microsecond. This information is valuable for determining the effect of wave reflections from objects.

THE NOISE PROBLEM

Television requires an extremely wide band to accommodate both the video modulation and the sound. One of the principles of radio communication states that when amplitude modulation is employed, the noise susceptibility of a system is proportional to its bandwidth. For example, automobile ignition systems produce interference. Because of the length of the spark-plug wiring, this interference is quite often tuned and radiated within the wide television bands. Fortunately, the strength of a radiated interfering field drops rapidly as the distance from the source of interference increases. If the tv antenna system is relatively high, is not sensitive to vertically polarized waves, and has a balanced lead-in system, the effects of this and other types of man-made interference can be reduced. These are additional reasons for the installation of an efficient antenna in weak-signal areas.

GHOSTS DUE TO MULTIPLE-PATH TRANSMISSION

Before the television antenna is considered in detail, let us examine some of the common picture distortions caused by indirect transmission paths and how these defects can be overcome or alleviated.

Fig. 11-8A shows a condition which might occur in an urban residential area. The direct signal arrives at the receiver over path A (assumed to be 10,000 feet). When the signal is 12,800 feet from the transmitter, the office building reflects the signal to the receiver over path C (9000 feet). The combined path B plus C is 21,800 feet. This reflected-wave path is 11,800 feet longer than the direct path. We find, from the wavelength-frequency relationship in Fig. 11-7, that the wave travels 984 feet per microsecond. In the foregoing instance, the reflected wave will arrive 11,800/984, or 12 microseconds, later than the direct wave.

Fig. 11-8B shows the effect of the delayed, or ghost, signal on the picture. The horizontal-scanning time relationships discussed in Chapter 7 will explain the image displacement. Approximately 54 microseconds elapse while one line is being scanned. On a twenty-one-inch tube, the picture is about nineteen inches wide. The signal arriving over path B plus C requires twelve microseconds more than the direct path signal and, during this time, scanning will have advanced four and one-quarter inches. Thus, a second image due to the reflected wave will appear displaced from the main image as shown in Fig. 11-8B. This ghost image

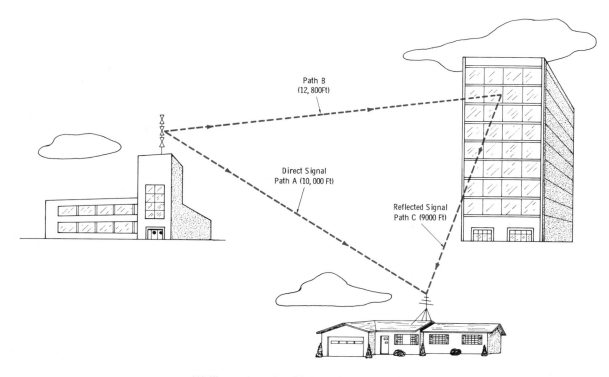

(A) Illustration of multiple-path transmission.

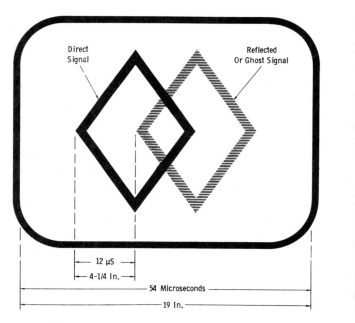

(B) Ghost caused by reflection over path B plus C.

(C) Photograph of ghost due to single-reflection path.

Fig. 11-8. Production of a ghost image by multiple-path transmission.

will generally suffer some loss of signal strength due to imperfect reflection and to energy absorption and, therefore, will be less intense than the image caused by the direct signal. Fig. 11-8C is a photograph of a ghost of the type just discussed.

In the example of Fig. 11-8, the ghost was caused by a single reflection. Often, a series of overlapping ghost images, due to multiple reflections from a number of buildings or from other obstructions, will appear. When such multiple ghost images arrive over paths only slightly longer than the direct-signal path, vertical lines on the picture are widened, and the fine detail is obscured. Since the picture appears smeared, this phenomenon is often referred to in television slang as a "smear ghost."

Ghost images are eliminated or reduced until they are no longer objectionable by using the directional properties of certain antennas. The theory and application of such antennas will be covered later in this section. Circular polarization, as mentioned previously, helps to reduce ghost images due to reflected signals.

In certain urban locations, the receiving antenna is shielded by buildings from the line-of-sight signal from the transmitter. One of the reflection paths will often produce a stronger signal than the direct path. Therefore, the antenna is beamed to accept the stronger reflected signal. The dominant reflection is then considered a substitute primary signal, and the direct-path signal produces a ghost which must be suppressed.

GHOSTS DUE TO REFLECTIONS IN THE LEAD-IN

The resonant antennas employed for television require special lead-ins. This lead-in (or transmission line) usually consists of either a parallel-wire line, a shielded parallel-wire line, or a coaxial (concentric) cable.

The reasons for these special types of lead-ins are: (1) to transfer the maximum amount of energy from the low-impedance antenna to the rf input of the receiver, and (2) to restrict signal pickup to the antenna only.

Maximum energy transfer of the intercepted signal from the antenna to the receiver input requires that the characteristic impedance of the line be matched to the impedances of the antenna and the receiver input circuit. The various types of lead-ins and their applications will be covered

in greater detail later. At present, we will consider the effect of line mismatch in the production of ghost images.

When the impedance of the transmission line differs considerably from that of the antenna or the receiver input, energy will be reflected from the receiver to the antenna and back to the receiver again. A ghost image is produced when the reflection reaches the receiver. Unless the mismatch is considerable, only one secondary or ghost image will be produced.

When ghost images are caused by line mismatch, only a small displacement of the image occurs because of the relatively short transmission line. If the lead-in is less than 70 feet (21.3 meters) long, the ghost image will be so close to the main image that the eye cannot distinguish it as a separate image. The horizontal resolution will be impaired, and the image will appear improperly focused.

It is possible to distinguish between a ghost image due to line reflection and one due to reflection by some object. If the antenna is rotated while watching the image, a ghost due to line mismatch will not change. On the other hand, if the ghost is due to a reflected path, its intensity will change with respect to the main image due to the directional characteristic of the antenna.

THE HALF-WAVE DIPOLE

Having reviewed the nature of wave propagation and some of the basic phenomena of the transmission of television signals, we are now prepared to examine the types of antennas which have been developed to cope with the problems encountered.

Since the advent of commercial television in the United States, there has been much activity in the search for types of antennas to meet requirements. Among the desired characteristics are more-uniform response or higher pickup and sharper directional patterns.

Practically all television antennas are variations of the basic half-wave dipole or Hertzian doublet. An understanding of this resonant type of antenna will help in the study of the more complicated arrays. The half-wave dipole is the industry standard against which the efficiency of other types of antennas is compared.

Fig. 11-9A shows the concept of a half-wave dipole in space. If such an antenna were of very small wire and could be removed from the earth's

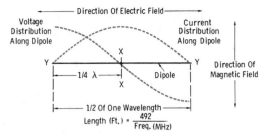

(A) A theoretical dipole (a very thin wire isolated in space).
Length equals half of carrier-wavelength.

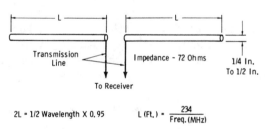

(B) Electrical circuit equivalent of dipole.

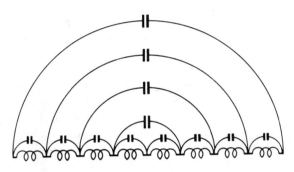

(C) A practical receiving dipole.

Fig. 11-9. The half-wave dipole antenna.

influence, its resonant electrical length would equal its physical length. This length, in turn, would be one-half the wavelength of the electromagnetic wave in space.

Although Fig. 11-1 shows the electric and magnetic fields as fixed uniform lines, both of these fields actually are varying at a sine-wave rate. Thus, a varying magnetic field will induce an alternating flow of electrons in the dipole wire, and a wave of current will pass down the dipole to its end. At the end, reflection will occur, and a standing wave having the voltage and current distributions shown in Fig. 11-9A will build up along the wire. Since the length of the theoretical dipole in Fig. 11-9A is equal to one-half wave in space, the voltage along the wire and the current through it will reach a resonant condition similar to that found in a parallel-tuned circuit. Actually, the

dipole may be considered such a tuned circuit, but here the inductance is distributed along the length of the wire, and the capacitance consists of many elements across the individual small inductances. This tuned-circuit equivalent is shown in Fig. 11-9B.

The foregoing discussion has been concerned with hypothetical dipoles. They could never exist because the dipole wire must be infinitely small in diameter compared with its length and the dipole must be far enough from surrounding objects, including the earth itself, that there would be no influence from such objects. In practice, the dipole must be erected within a few wavelengths of ground and must be large enough in diameter to be self-supporting. Both requirements cause the velocity of wave travel in the practical dipole to be less than the speed of electromagnetic waves in free space.

Although the length of the dipole for resonance will vary slightly depending on the proximity of other objects, an average figure for the reduced length due to the proximity effect is 5%. In other words, the half-wavelength value obtained from the formulas in Fig. 11-7 should be multiplied by 0.95 to find the length of a dipole for any given frequency. The formula for a practical dipole is found in Fig. 11-9C.

The dipole provides the greatest amount of signal pickup when parallel to the electric field of the wave front. In other words, a line drawn from the receiving antenna to the transmitter should be at right angles to the length of the dipole. As the dipole is rotated, its response varies as shown in Fig. 11-10A. A curve of this type is known as a polar response curve. The response at different angles away from the normal or maximum-response position is determined by taking the ratio of the length of the radius at any point to the maximum radius.

The response curve of Fig. 11-10A represents a cross section taken in the plane of the antenna. The response to waves arriving from the sky or reflected from the ground is determined by the solid figure obtained by rotating the response curve in Fig. 11-10A around axis X-Y, as shown in Fig. 11-10B. The resulting solid figure is doughnut shaped.

From the reception pattern in Fig. 11-10A, we see that the simple dipole is bidirectional and receives equally well from front or rear, but it shows practically no response in line with the length of

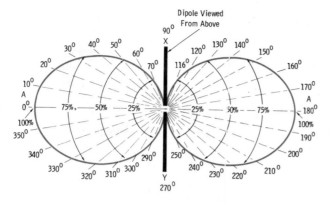

(A) Reception pattern of a dipole showing relative response versus direction in a horizontal plane.

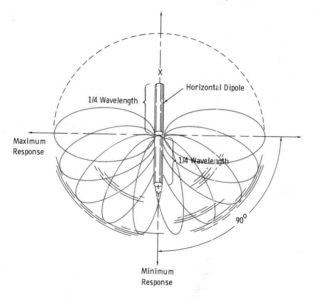

(B) Solid pattern in space of half-wave dipole.

Fig. 11-10. The reception pattern in space of a half-wave dipole antenna.

the dipole. This directional characteristic is valuable for discriminating against spurious reflections, which would produce ghosts in the picture. The simple dipole is satisfactory when the signal strength is high, when only one station (or stations whose frequencies are close to each other) is to be received, and when the ghost problem is not too severe.

If greater directional selectivity or more signal pickup is required, the pattern and response of the simple dipole can be improved by additional elements known as reflectors and directors. The use of these elements in more complicated antenna structures or arrays will be covered in detail later.

The impedance of the simple half-wave dipole at its center is approximately 72 ohms. For maximum energy transfer, the lead-in also should have a characteristic impedance of 72 ohms. This impedance value is rather low and is best suited to the coaxial-type lead-in. The mismatch of the antenna to the lead-in can vary as much as two to one without serious signal loss and without ghosts, if the line is appropriately matched to the receiver input.

The simple half-wave dipole is most efficient when its length is correct for a particular carrier frequency. If several adjacent television channels are to be received with a simple antenna, its length should be made correct for the average wavelength of the lowest and highest channel frequencies received. For example, if a station on Channel 3 (video carrier frequency, 61.25 MHz) and a station on Channel 6 (video carrier frequency, 83.25 MHz) are to be received with approximately equal antenna response, the dipole should be made the correct length for the average wavelength of these frequencies. Therefore, if the proper length of a dipole to receive Channel 3 is 7'7¾" (2.33 meters) and the proper length for Channel 6 is 5'7½" (1.71 meters), then the length required to give an approximately even response over both channels with a single dipole would be the average length or the sum of both divided by two. For example:

$$7'7\tfrac{3}{4}'' + 5'7\tfrac{1}{2}'' = 13'3\tfrac{1}{4}''$$
$$= 159\tfrac{1}{4}''$$

Then

$$159\tfrac{1}{4}'' \div 2 = 79\tfrac{5}{8}'' \ (2.02 \text{ meters})$$

This would be the overall length of the dipole for a fairly equal response between the two channels. The length of each dipole section would thus be:

$$79\tfrac{5}{8}'' \div 2 = 39\tfrac{13}{16}'' \ (1.01 \text{ meters})$$

THE FOLDED DIPOLE

If two half-wave dipoles are placed parallel to each other with their ends connected, the received currents will be in phase. The reaction of one dipole on the other will increase the impedance at the center of the dipole from 72 ohms to approximately 300 ohms. Fig. 11-11 shows such a folded dipole and the formula for calculating the re-

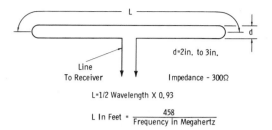

Fig. 11-11. Dimensions of a folded dipole.

quired dimensions. The folded dipole possesses several advantages over the single half-wave dipole.

1. Its higher impedance provides an ideal match for the popular parallel, or twin-lead, transmission line. This type of television lead-in has a characteristic impedance of 300 ohms which is the input impedance of nearly all present-day television receivers (some receivers also have a 72-ohm input for coaxial cable).
2. It is more receptive to a wider frequency band than the simple half-wave dipole, but has the same directional pattern.
3. Because of its design, it has a more rigid structure and will withstand greater wind pressure.

ANTENNA STRUCTURES EMPLOYING THE DIPOLE WITH REFLECTORS AND/OR DIRECTORS—YAGI ARRAYS

If a second half-wave dipole is placed parallel to and closer than a half-wave length from the receiving dipole, it is called a "parasitic" element. The magnetic and electrostatic fields of the parasitic element will modify the directional pattern and increase the gain of the receiving dipole. When a parasitic element is on the side away from the transmitter, the element is known as a reflector. An element on the side toward the transmitter is known as a director. A reflector is approximately 5% larger than the receiving antenna, and a director is about 4% shorter. This is equivalent to saying that the reflector is tuned to a somewhat lower frequency than the operating frequency, and that the director is tuned to a slightly higher frequency. Fig. 11-12A shows the arrangement of the dipole with a director and a reflector. The effect of the spacing of these ele-

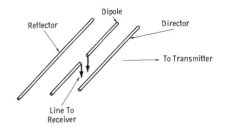

(A) Dipole with director and reflector.

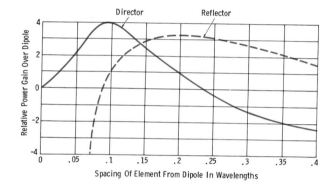

(B) Effect of director and reflector spacing on gain of parasitic array.

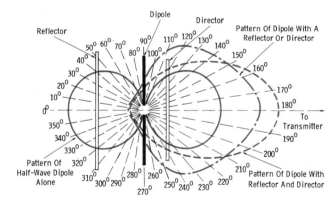

(C) Typical reception patterns of a dipole with director and reflector elements singly or in combination.

Fig. 11-12. The dipole antenna with parasitic elements (directors and reflectors).

ments on the power gain of the dipole antenna is shown in Fig. 11-12B.

Fig. 11-12C compares the directional pattern and gain of the dipole alone with a dipole using either a reflector or a director, or both. The gain is increased by these additional elements, and the directional pattern is made sharper, so that stations at other points of the compass are received with less signal strength. For this reason, arrays

Fig. 11-13. Antenna with three separate arrays, each designed to receive different channels.

THE BROADBAND PROBLEM

The half-wave dipole and the arrays employing half-wave elements are most efficient at the frequency for which they have been cut. Satisfactory performance is obtained only when one station serves the area or when relatively high signal strength exists for several stations in the same television band. When stations in both the low vhf band (54 MHz to 88 MHz) and the high vhf band (174 MHz to 216 MHz) must be received, antennas with wider frequency response are necessary. Of course, even wider response is required when the antenna must receive stations in the uhf band (470 MHz to 890 MHz), in addition to those in the vhf bands.

Antennas employing separate arrays for each channel, as shown in Fig. 11-13, have become popular in many metropolitan areas. Using a separate dipole-element array for each channel overcomes the narrow-band limitations of a single dipole. Also, the use of separate arrays eliminates directional problems encountered when the transmitters are located in different directions from the receiving antenna.

The folded dipole exhibits the broadest frequency response of any of the television antennas discussed up to this point. Fig. 11-14 shows a combination of folded dipoles which provides directional discrimination and reception in both bands. This arrangement is called an "in-line" antenna. In the low vhf band, the larger folded dipole backed by a reflector is the receiving antenna, and the smaller folded dipole acts as a director. In the high vhf bands, the small folded dipole functions as the antenna, and the large folded dipole behind it acts as a reflector. The two dipoles are connected to each other at their center (high-current) points

of this type will increase the pickup from the desired station and, at the same time, suppress reflection paths which would produce ghosts. Multiple-element arrays such as this are known as *Yagis*, after the Japanese physicist Hidetsugu Yagi who originated them.

Arrays with directors and reflectors, although an improvement over the dipole from the standpoint of pattern and gain, will not accept as wide a frequency band as the simple dipole or folded dipole. It is often necessary to erect a number of these antennas tuned to various stations in the band. Fig. 11-13 shows three sets of arrays on a single mast. The lower array is for Channel 4, the center array is for Channel 8, and the upper array is for Channels 6 and 13. Antennas with as many as four arrays, each designed for a single channel, are used in some areas.

If increased gain and directivity are required, additional director elements can be added. Arrays with as many as ten elements are often employed to receive a single station when the receiving location is in a fringe area.

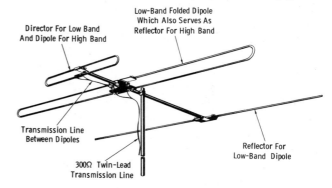

Fig. 11-14. "In-line" antenna array of folded dipole for reception on both vhf bands.

by the proper length of twin-lead transmission line so that a transition occurs between the two bands. An additional feature of the array, determined by the size of the elements and their spacing, is a low efficiency in the gap between the two vhf bands. This gap contains the fm broadcast stations (88 MHz to 108 MHz) which may be a source of television interference. In the following text, means for increasing the bandwidth, other than by the use of the folded dipole, will be discussed.

A large-diameter dipole will have a wider frequency response than a smaller-diameter dipole of the same length. If the dipole is a large cylinder, i.e., three to six inches in diameter, its Q will be decreased, and its response will be broadened. Such a cylinder would be awkward to install and could easily be damaged by the wind. An equivalent of the cylinder can be obtained by constructing a cage of wires with their ends connected to rings.

The cage construction can be further modified to the form of two cones whose apexes meet at the lead-in or feed point. Fig. 11-15 shows such an antenna. However, this is still not a practical design because of its awkward construction. By taking a cross section of the antenna of Fig. 11-15, antenna designers arrived at the fan-type antenna of Fig. 11-16. This design is known as a "conical" antenna and has a broad frequency response.

The V-type antenna consists of a dipole with V-shaped elements. Moving the dipole elements from a straight line broadens the response, yet retains the directional pattern. Fig. 11-17 shows

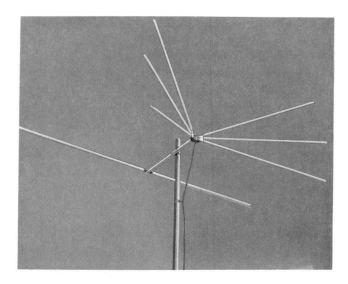

Fig. 11-16. A conical antenna having a broadband response pattern.

a simple V antenna, commonly referred to as "rabbit ears." These antennas are used indoors for reception of local stations.

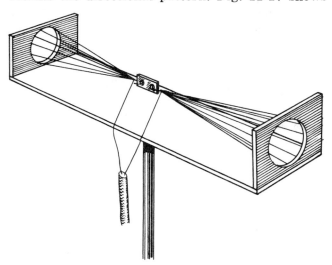

Fig. 11-15. The "cage" antenna is a variation of the dipole for wideband reception.

Fig. 11-17. A simple V-type antenna, commonly known as "rabbit ears," used indoors for local reception.

STACKED ARRAYS

Most types of antennas can be stacked by erecting another identical antenna above the first in the same vertical plane. The antennas are critically spaced (usually one-half wavelength apart) to provide in-phase operation to a common lead-in. The advantages of vertical stacking are:

1. Additional gain is obtained because of the added antenna.
2. Some vertical directivity is contributed by the mutual interaction of the antennas. This interaction discriminates against a reflected wave from the ground and confines the reception to the direct, or sky, wave. Fig. 11-18 illustrates several examples of vertically stacked arrays.

THE CORNER-REFLECTOR ANTENNA

Fig. 11-19 shows an antenna structure which, because of its very high front-to-back ratio, greatly increases the pickup of the dipole. The use of this type of television antenna is limited to the uhf range, since the reflector would be too large to be practical for the vhf bands. This type of antenna is used for uhf reception in fringe areas.

The graph in Fig. 11-19B illustrates the ratio of the power received with the dipole at various spacings from the reflector, to the power received with the dipole alone. The three curves represent the power ratios for reflector corner angles of 90°, 60°, and 45°. Fig. 11-19C illustrates the influence of the corner reflector on the impedance of the antenna. Since the corner reflector can be used with

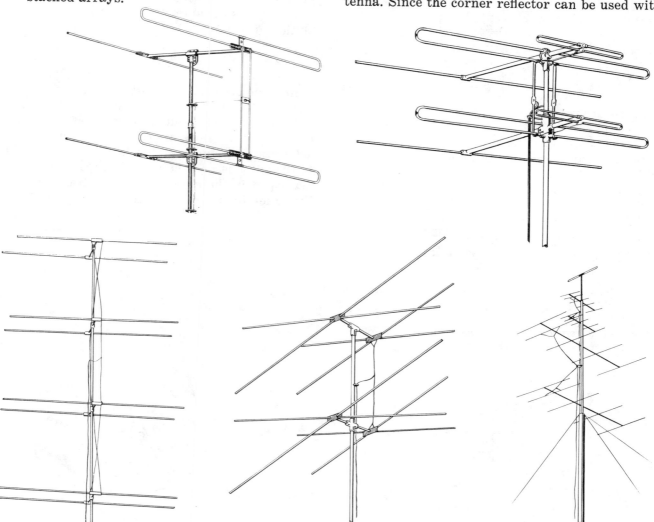

Fig. 11-18. Stacked arrays for higher gain and improved directional characteristics.

117

any type of dipole, the change in impedance is represented as a percentage.

ANTENNA ROTATORS

Since most television antennas are directional, the receiving antenna should be pointed toward the transmitting antenna for best reception. This creates a problem in areas where television signals are received from more than one direction. Also, in areas where there are multiple-path re-

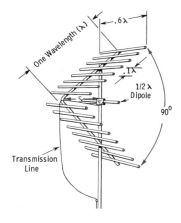

(A) 90° corner-reflector antenna.

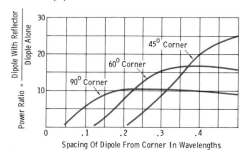

(B) Effect of corner angle and dipole position on gain of corner reflector.

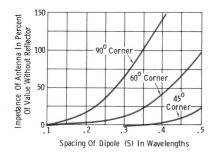

(C) Effect of corner angle and spacing on dipole impedance.

Fig. 11-19. The corner-reflector antenna.

ception problems, the antenna must be pointed in the direction that provides the best signal. Therefore, for best reception on each channel, antenna rotators are employed.

Fig. 11-20 shows a remote-controlled antenna rotator. The motor which rotates the antenna is mounted on the antenna mast and the antenna is mounted to the motor. The control unit for the rotator is located at the receiver, and power is fed to the motor by a cable from the control unit. Most control units indicate the direction in which the antenna is pointing. This enables the user to record the best position of the antenna for any given station and to subsequently return the antenna to this position. The rotator motor is geared for relatively slow rotation of the antenna so that the point of best reception will not be overlooked. The rotator turns the antenna through a complete revolution of 360°.

TYPES OF LEAD-IN

In our discussion of ghosts produced by reflections in lead-in cables (transmission lines), we indicated that maximum power transfer and freedom from reflections are obtained when the characteristic impedance of the lead-in matches that of the antenna and the receiver input. The power developed in the antenna must be transferred to the tuner input with as little loss as possible in

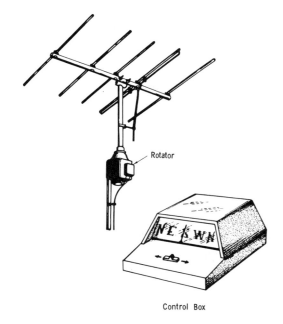

Fig. 11-20. Typical antenna rotator installation.

order to override noise and produce a steady, high-quality picture.

Two basic types of transmission line (or lead-in) are used for tv receivers. These are the two-wire parallel line and the coaxial or concentric-cable line. The types of two-wire parallel line can be further subdivided into the flat-ribbon, solid-polyethylene dielectric type; the tubular-polyethylene type, with either a hollow or cellular-polyethylene core; and the oval-shaped line, encapsulated in cellular polyethylene (either shielded or unshielded). These lead-ins are illustrated in Fig. 11-21. Each of these types exhibits its own particular characteristics of impedance and of loss, or signal attenuation.

Two-Wire Parallel Lines

Flat-Ribbon Parallel Line—Many technicians have used the flat-ribbon lead-in almost exclusively in the past. Although many examples of satisfactory reception using this type of lead-in can be cited, flat line deteriorates when exposed to the elements.

Modern lead-in must carry both uhf and vhf signals, including the critical color signals. Any installation which requires an outside antenna should use a lead-in which is specifically designed

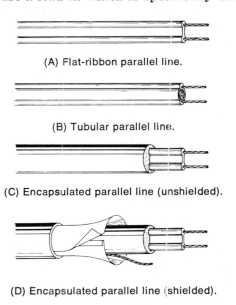

(A) Flat-ribbon parallel line.

(B) Tubular parallel line.

(C) Encapsulated parallel line (unshielded).

(D) Encapsulated parallel line (shielded).

(E) Coaxial cable.

Fig. 11-21. Types of television-antenna transmission line or "lead in."

for outdoor operation. When moisture accumulates on the surface of a flat line, losses occur due to the partial shorting effect of moisture and dirt on the rf field surrounding the conductors. When the flat line is routed close to conductive objects, part of the electromagnetic field surrounding the lead-in is induced into these objects, increasing attenuation losses.

Tubular Parallel Line—There are two types of parallel line which are of the tubular construction. In the first type, the solid-polyethylene tube has a hollow core, while the second type has a cellular-polyethylene core. Cellular-core lead-in is designed to reduce wet-weather effects and normal dielectric losses by providing a more-uniform, and relatively low, dielectric-constant (1.5) material between the conductors. Hollow tubular lines, because the dielectric between the wires is mostly air, have a lower dielectric constant (1.0 for dry air) in good weather conditions. However, moisture tends to accumulate inside the hollow tube through condensation caused by temperature changes. This causes the impedance between the conductors to drop considerably, which, in turn, results in greatly increased signal losses.

The cellular-core line has an additional advantage over the hollow-core type in that it is more rigid, making it less susceptible to flutter in heavy winds. In addition, the use of cellular-core line eliminates the need for sealing the end of the tube after installation. The cellular-polyethylene core also eliminates moisture condensation.

Encapsulated Parallel Line—A more-recent development in outdoor tv transmission line for home receivers is the encapsulated line. This line has the two conductors fully encased in an oval-shaped cellular-polyethylene jacket. This keeps surface deposits out of the critical signal areas under any weather conditions.

The introduction of encapsulated lead-in (both shielded and unshielded) has brought about a minimization of signal losses due to proximity effects. A major portion of the electromagnetic field surrounding the two conductors is within the encapsulating cellular outer covering. This thick covering acts as a barrier to nearby conductive surfaces into which the field would otherwise penetrate, seriously reducing signal strength and picture clarity.

When using the unshielded type of encapsulated lead-in, the usual practice of using stand-off insulators a minimum of six inches long, spaced every four to six feet, and twisting the cable about every 18 inches should still be observed. (Stand-offs which do not encircle the lead-in with metal are required for uhf.) However, the use of this type of lead-in will still produce a better-quality picture than the older type of transmission line, even in weak signal areas, because of the minimization of proximity effects. This is provided, of course, that the line is being used in rural areas or any other relatively noise-free area. In unusually noisy locations (i.e., areas where man-made electrical interference, such as automobile ignition noise, 60-Hz power-line interference, etc., is unusually severe), the lead-in should also be shielded. Shielded twin-lead has been available for some time, but only in lower impedances, such as 40 or 100 ohms. Consequently, the losses per foot for this type of line were usually higher than coaxial cable of the same impedance. The primary advantages of the older shielded twin lead were that it reduced pick-up of man-made electrical noise and, being a balanced line, did not require matching transformers at both the antenna and the receiver terminations. Also, shielded twin lead eliminates out-of-phase signal pick-up caused by the line acting as an antenna.

Recent developments in shielding materials have now made it possible to provide the newer 300-ohm encapsulated twin-lead in shielded form, without adding greatly to the overall cable dimensions. The new shielding consists of a thin aluminum foil bonded to a Mylar tape insulator. This laminated tape is then wound spirally around the encapsulated twin lead. A ground or "drain" wire, which is in intimate contact with the aluminum side of the tape throughout the length of the lead-in, is provided for easy ground termination. Finally, the entire cable is encased in a waterproof polyethylene outer jacket, making this the first truly dependable all-weather 300-ohm line. This line does not require stand-off insulators, twisting, or careful routing to minimize proximity effects. It may be taped directly to the antenna mast or tower, routed through metal conduit, buried underground, or even installed in rain-filled gutters.

Shielded twin-lead, unlike coaxial cable, does not require a copper-wire braided shield to carry current, because the shield of any two-conductor transmission line is not in the signal circuit. Thus, the new Mylar/aluminum-foil shield, thinner than braided wire, can be used with the new 300-ohm twin-lead without adding appreciably to the overall dimensions of the cable. In addition, this type of shield provides 100% shielding coverage, whereas braided wire can offer only about 95% coverage, since there are minute gaps or holes between the wires. This is very important at rf frequencies, especially those in the tv range. Even at uhf frequencies, a 100-foot run of the shielded twin-lead will deliver approximately 50% of the antenna signal to the tv receiver. This compares very favorably to coaxial cable, a 100-foot run of which will deliver only about 15 to 25% of a uhf antenna signal to the receiver.

Coaxial Cable

Coaxial cable (coax) for television lead-in consists of a flexible conductor molded in the center of a solid (or cellular) polyethylene cylinder. This cylinder is surrounded by a braided copper-wire outer conductor, and the entire cable is covered by a weatherproof vinyl or polyethylene jacket. The outer conductor is grounded at the receiver and acts as a shield for the inner conductor.

Coaxial cable exhibits low loss, is free from noise pickup, and is available with characteristic impedances from 50 to 150 ohms. For television antenna installations using coaxial cable, the 75-ohm impedance is used almost exclusively. With the rapidly increasing growth of new uhf stations since 1964, when the FCC made it mandatory for all tv receiver manufacturers to incorporate all-channel tuners in their sets, the use of 75-ohm coax for outdoor purposes has increased considerably. Although it has somewhat higher losses per foot than the new encapsulated and shielded 300-ohm twin lead, coax is somewhat less expensive and will provide reliable, noise-free signals in troublesome signal areas. Coax does, however, require the use of matching transformers at both the antenna and receiver ends, since most receivers and antennas have 300-ohm terminal impedances. However, in recent years some antenna manufacturers, as well as receiver manufacturers, have begun to supply their products with 75-ohm terminations.

The characteristic impedance of coaxial cable is determined by the ratio of the diameter of the outer conductor to the diameter of the inner conductor. For this reason, impedances higher than 150 ohms would require either a large diameter outer conductor or an extremely small and weak inner conductor. Fig. 11-21E shows a sample of 75-ohm coaxial cable for television use.

TELEVISION RECEPTION IN FRINGE AREAS

The service area of a television transmitter for rural and residential areas is normally defined by a contour line beyond which the signal strength falls below about 500 microvolts per meter. This contour depends on the power of the transmitter, the height of the transmitting antenna, the topography of the land, and the effect of shielding by buildings and other structures.

Fringe reception at points beyond the service area requires highly efficient antennas. Yagi antennas, like the one in Fig. 11-22, are suitable for the reception of individual channels, and modified Yagis can be used for multichannel reception. Fig. 11-23 shows a popular wideband antenna for all-channel reception. As shown in Fig. 11-5, the vhf range is increased by additional height at the

Fig. 11-22. Multielement yagi antenna.

Fig. 11-23. Broadband yagi antenna for medium-fringe reception on all vhf channels.

receiving location. Towers for this purpose are commercially available. Antenna amplifiers or boosters can provide additional rf gain, and many types of these are also used in weak signal areas.

Further antenna research in recent years has led to the development of a new concept in broadband, high-gain antenna design. Known as log-periodic spacing, this design is based on the "transposed-harness" type of antenna, which incorporates a large number of active dipoles connected together in a special phasing arrangement. This results in reinforcement of signals arriving from the front of the array and cancellation of pickup from the rear.

The log-periodic concept spaces the active dipoles in accordance with a mathematical formula (logarithmic) based on the theory of an infinite spiral. Such spacing broadens the bandwidth of the array by blending the characteristics of each dipole into a uniform frequency response that covers a very wide range of television frequencies. Early log-periodic antennas were designed to cover just the two vhf bands, while more recent developments have provided an antenna with a bandwidth capable of covering the entire range of tv frequencies from vhf Channel 2 through uhf Channel 83. An antenna of this type is shown in Fig. 11-24.

An even more recent innovation has been the color-laser, log-periodic antenna. Developed primarily for color reception, because of the more stringent requirements of color tv, this antenna differs from the basic log-periodic design in that the thin dipoles of the driver and director elements are replaced by ones of much greater cross-

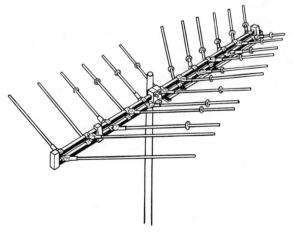

Fig. 11-24. Log-periodic antenna providing coverage on all tv bands (vhf and uhf).

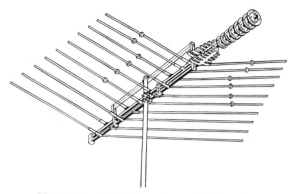

Fig. 11-25. A color-laser log-periodic antenna.

section. This increases the bandwidth of each individual element. Also, the log-periodic driver is modified to tailor its gain characteristic for use with broadband directors.

The directors of the color-laser antenna are a result of radar antenna design. The directors are circular and, as can be seen in Fig. 11-25, resemble a collection of discs on a rod.

MASTER ANTENNA SYSTEMS

A master antenna system (matv) is an installation which allows several television receivers to be fed from a single antenna. In recent years, such systems have come into widespread use in apartment buildings, hotels and motels, schools, hospitals, and even private homes. A simple matv system, such as would be used in a home, is shown in Fig. 11-26. Of course, a system used in an apartment building or a motel would be more complex and would require additional splitters, distribution amplifiers, extender amplifiers for long cable runs, and other components.

The antenna chosen for a matv system depends on factors such as: the location of the system, the

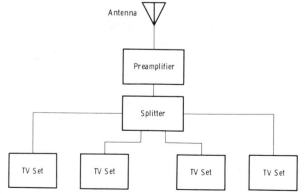

Fig. 11-26. Block diagram of a simple master antenna system.

direction and operating channels of the stations to be received, and the signal strength required at the input to the preamplifier. If the system is located in an urban area where the signal is strong, a simple Yagi antenna may be sufficient. In a fringe area, an antenna with high gain will be necessary. When the stations to be received are located in different directions, an array with an antenna for each channel may be employed. Many systems use a separate antenna for each channel and feed the signals into a common distribution amplifier. In a remote fringe area, where the stations to be received are in the same direction from the antenna, a wideband, high-gain antenna such as the log periodic may be used. The signal required from the antenna will depend on the gain of the system and the signal required at various points within the system. The system is usually designed so that the signal strength from each channel is about the same at the receiver.

QUESTIONS

1. When the electrostatic lines of force of a television signal are horizontal to the earth's surface, is the signal horizontally or vertically polarized? What type of polarization is standard for telecasting in the United States?

2. What determines the transfer of maximum energy of the signal from the antenna to the receiver input?

3. How are ghost images due to line reflections differentiated from those due to signal reflections?

4. At what position, with respect to the transmitted signal, will a receiving dipole produce the greatest current?

5. What is the characteristic impedance of a half-wave dipole? A folded dipole?

6. What are the director and reflector elements of an antenna assembly? How are they identified?

7. What is meant by stacked arrays? What are their advantages?

EXERCISES

1. Sketch the following types of antennas:
    (a) A single dipole.
    (b) A folded dipole.
    (c) A dipole with a director and a reflector.

2. Calculate the length of a dipole element to be used for Channel 4 and one to be used for Channel 8.

Tuners

The tuner of a television receiver consists, basically, of an rf amplifier, a local oscillator, and a mixer stage. This combination of circuit elements performs the same function as its counterpart in a conventional superheterodyne radio receiver. However, a number of complications not encountered in the reception of ordinary broadcast signals require a more complicated design than is found in standard broadcast receivers.

The broadband nature of the television signal requires that the tuner accept and amplify a band of frequencies 6-MHz wide. The frequency allocation of television channels (54 MHz to 88 MHz and 174 MHz to 216 MHz for the vhf band, and 470 Hz to 890 MHz for the uhf band) necessitates special types of coupling circuits to maintain uniform gain at the extremes of each band. Balanced types of input circuits used in television tuners must be matched to the impedance characteristics of available transmission line. The tuner must also be designed to reject undesired or spurious signals outside the desired television band. These spurious signals may be caused by:

1. The adjacent-channel sound carrier.
2. Cross modulation due to other television channels.
3. Direct transmission of signals at the intermediate frequency through the rf system.
4. Interferences due to other television and fm signals.

Another consideration in television tuner design is the energy radiated by the local oscillator. Such radiation must be suppressed to prevent interference with neighboring television receivers.

TUNING SYSTEMS

A wide variety of tuning systems has been employed in television tuners. Some of these systems are no longer used, while others are still used but in modified form. In this section, we will describe briefly some of the systems used in present-day vacuum-tube and solid-state television receivers.

Fig. 12-1 shows an example of a switch-type vhf tuner in which rotary wafer switches select the proper inductances for tuning to the desired channel. The schematic for a switch-type tuner is shown in Fig. 12-2. The tuner schematic shows a single-ended input for a 75-ohm coaxial cable. The antenna input circuit on the back of the receiver includes a balun transformer for matching a 300-ohm balanced line to the 75-ohm tuner input. A coaxial-cable fitting and a shorting link allow the user to select either the 75-ohm or the 300-ohm antenna input.

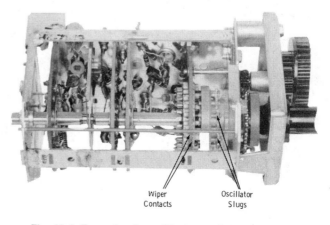

Fig. 12-1. Example of a solid-state switch-type tuner.

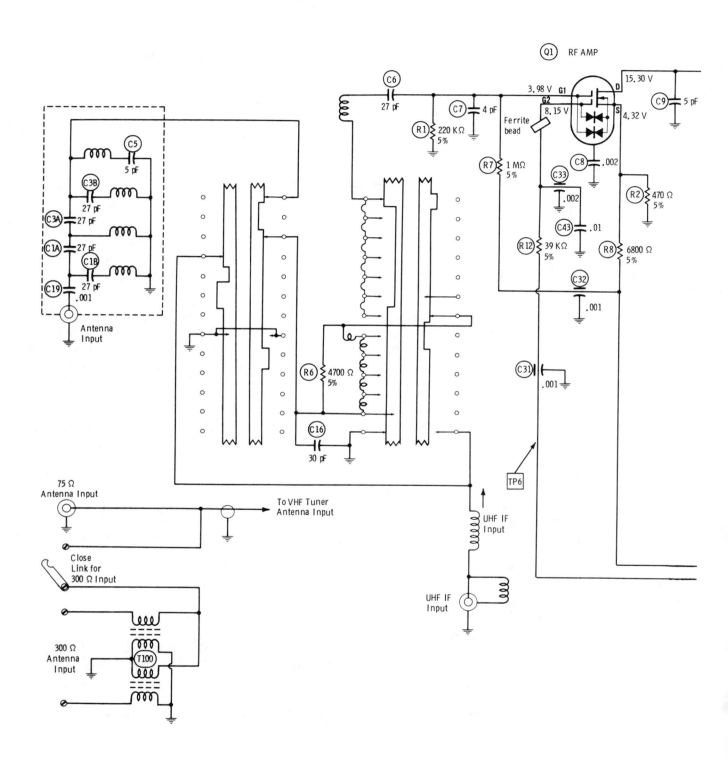

Fig. 12-2. Schematic diagram

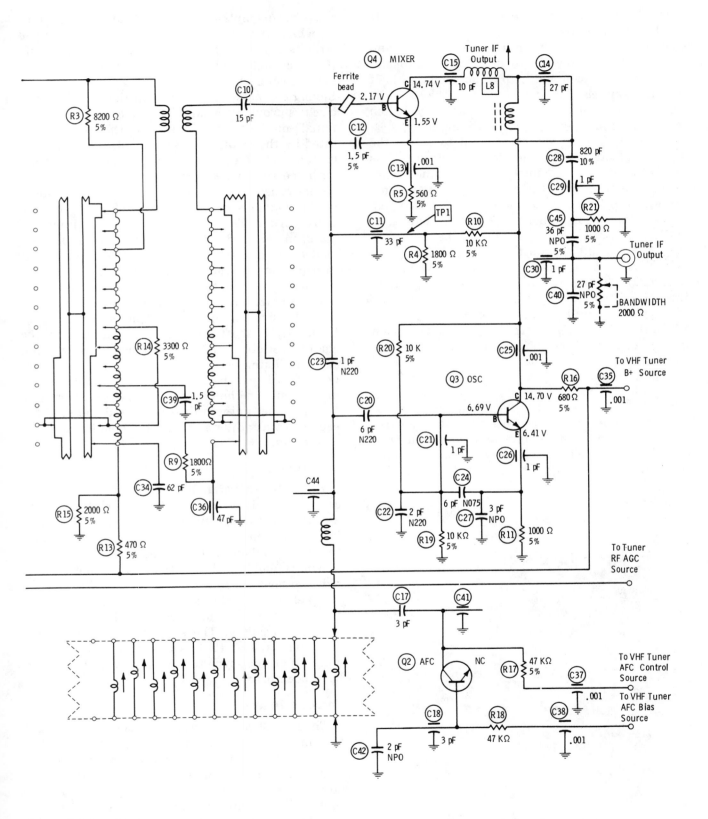

for a switch-type tuner.

The inductances in the drain and gate circuits of the rf amplifier are selected by the rotary switch. Interstage coupling between the rf amplifier and the mixer is provided by a transformer with sufficient bandwidth to accommodate all of the vhf channels. The inductances in the base circuit of the mixer are selected by a rotary switch in the same manner as for the rf amplifier.

A special rotary switch is used to select the proper inductances for the oscillator circuit. This switch allows each channel to be fine tuned separately and eliminates the need to adjust the fine-tuning control each time the channel is changed. The oscillator coil and slug for each channel is mounted on the rotor portion of the oscillator switch. Each coil has a set of wiper blades which mate with a set of stationary contacts when that particular channel has been selected. A gear on the end of the slug screw engages a nylon gear when the fine-tuning control is being adjusted. When the fine-tuning control is released, the nylon gear disengages from the gear on the slug screw. In this manner each channel can be fine tuned independently.

Turret Tuners

Turret tuners have been used in television receivers for quite some time and, although they have been reduced considerably in size, are still found in many tube-type sets. Figs. 12-3 and 12-4

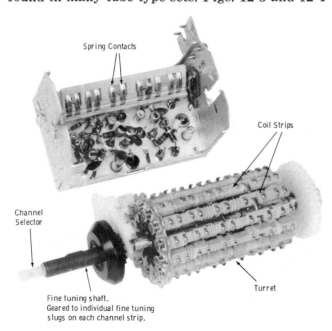

Fig. 12-3. A modern turret-type tuner.

show an example of a modern turret-type tuner in which the input, mixer, and oscillator tuned circuits are mounted on a rotating turret. Only the circuit elements for a single television channel are connected in the circuit at any one time. Spring contacts associated with the various tuner circuits provide a means for connection to the terminals of the tuned circuits. As the turret is rotated by the channel selector, contacts at the end of each set of tuned coils make positive contact with the stationary spring contacts.

The tuned coils for each channel are mounted on individual strips which are clipped into the turret drum. These strips contain the input, rf plate, mixer grid, and oscillator coils. A spring-loaded mechanism and a detent lock the turret into position at the desired channel. The spring contacts are mounted on an insulated strip and make contact with coil strips mounted in the turret. The fine-tuning mechanism is usually controlled by a shaft which is concentric with the channel-selector shaft. In modern turret tuners, each oscillator coil has a slug which is usually geared so it can be controlled by the fine-tuning mechanism. In the older turret tuners, there was a single fine-tuning adjustment, and the slug for each coil was adjusted through a hole in the front of the tuner.

Disc Tuners

Fig. 12-5 shows a disc tuner which combines certain principles of both the switch-type tuner and the turret-type tuner. Attached to the shaft inside the tuner are three large discs on which the tuning inductors are arranged. A detent mechanism, resembling that of the turret tuner, is provided to hold the rotating assembly in the desired channel position. Contact buttons and wipers, like those of the turret tuner, connect the discs to the external circuitry. The schematic for this tuner is shown in Fig. 12-6. Although this is a solid-state tuner, disc-type tuners employing tubes also have been used.

Varactor Tuners

The newest concept in tuner design is the varactor tuner. The most unique feature of the varactor tuner is the complete elimination of mechanical tuning devices within the tuner. Since there are no moving parts (such as wafer switches, turret contacts, and tuning shafts) in the varactor tuner, its dependability is greatly increased. From the

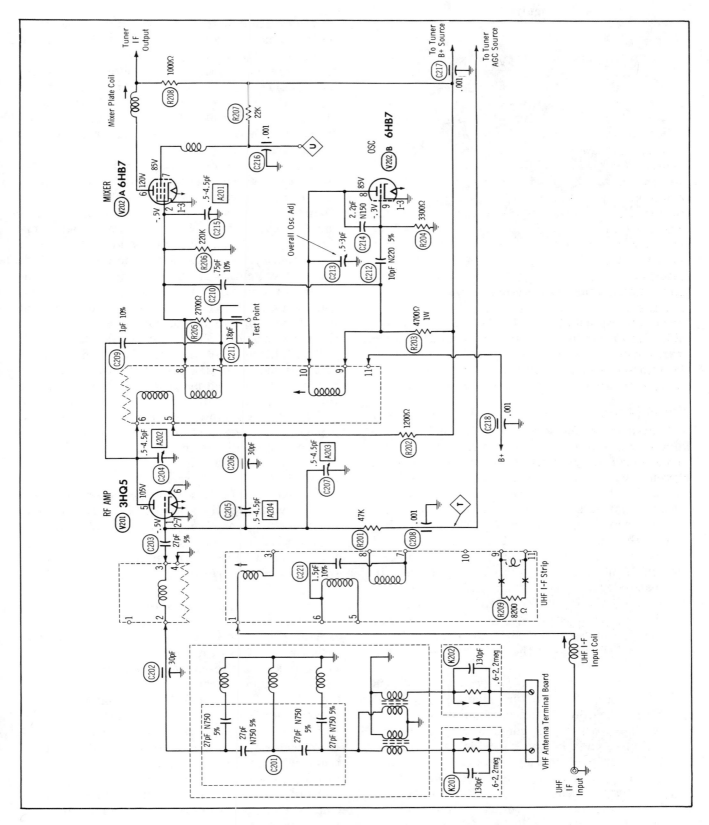

Fig. 12-4. Schematic of turret-type tuner shown in Fig. 12-3.

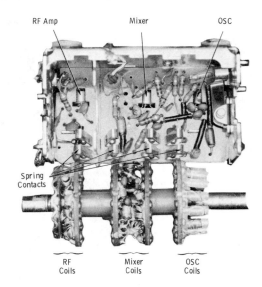

Fig. 12-5. Example of a modern disc-type tuner.

servicing standpoint, the absence of switch contacts eliminates the need for frequent and time-consuming tuner cleaning. Also, without the channel-selector and fine-tuning shafts to contend with, the tuner can be mounted almost anywhere that is convenient.

Tuning is accomplished in the varactor tuner by utilizing the characteristics of a special diode known as a varactor diode. When a diode is reverse-biased (Fig. 12-7), the electrons in the n-type material are attracted away from the junction of the diode. In the same manner, the holes (positive carriers) in the p-type material are attracted away from the junction. The area, adjacent to the junction, which has been depleted of carriers is known as the space-charge region or the depletion layer. Since there are no carriers in the space-charge region, this area is effectively a dielectric and the reverse-biased diode assumes the characteristics of a capacitor. The amount of reverse-bias voltage applied to the diode will determine the thickness of the dielectric region and, therefore, the capacitance of the diode.

The schematic of a typical varactor tuner is shown in Fig. 12-8. Varactor diodes are employed for tuning the antenna-input, rf-amplifier, mixer, and oscillator circuits. A common control voltage is applied to each of the varactors and the desired channel is selected by changing this control voltage. A rotary switch or a group of push-button switches may be used for applying the proper voltage to the varactors. The voltage supplied to the varactors by a given switch position is determined

by a separate potentiometer for each position (or push button). To set a given switch position to a particular channel, the potentiometer is adjusted to provide the proper control voltage for the varactors. The control switch for the tuner shown in Fig. 12-8 has twenty positions, with twelve for vhf and eight for uhf. Tuning for the uhf tuner is accomplished in the same manner as for the vhf tuner.

Since there is a rather large gap (86 MHz) between the two vhf bands, it would be impractical to tune the varactor tuner continuously from Channel 2 to Channel 13. Therefore, it is necessary to have a bandswitching arrangement in order to switch from one vhf band to the other. If conventional bandswitching methods with mechanical switches were used, many of the advantages of the varactor tuner would be lost. To accomplish bandswitching, diodes are used to switch inductances in and out of the tuned circuits. In the tuner shown in Fig. 12-8, diodes CR1, CR2, CR3, CR4, and CR6 are used for bandswitching in the tuned circuits. Diodes CR5, CR7, and CR8 are used for switching from vhf tuning to uhf tuning.

When a high-band channel has been selected, a positive voltage is applied to the anodes of the bandswitching diodes in the tuned circuits. This forward biases the diodes and causes them to conduct, effectively decreasing the inductances in the tuned circuits. Low-band operation applies a positive voltage to the cathodes of the diodes, causing them to become reverse-biased. With the diodes cut off, the inductance of the tuned circuit is increased and the resonant frequency is within the low-band vhf range.

RF AMPLIFIERS

Although most of the signal gain for a television receiver is provided by the if amplifiers, it is still necessary to employ an rf amplifier stage in the vhf tuner. Besides its primary function of amplifying the rf signal, the rf amplifier reduces cross modulation and improves the signal-to-noise ratio. Also, the rf amplifier acts as a buffer to prevent local-oscillator energy from reaching the antenna where it could be radiated as interference to other nearby television receivers.

All of the television tuners now being produced employ transistors and other solid-state devices. However, there are many television receivers with

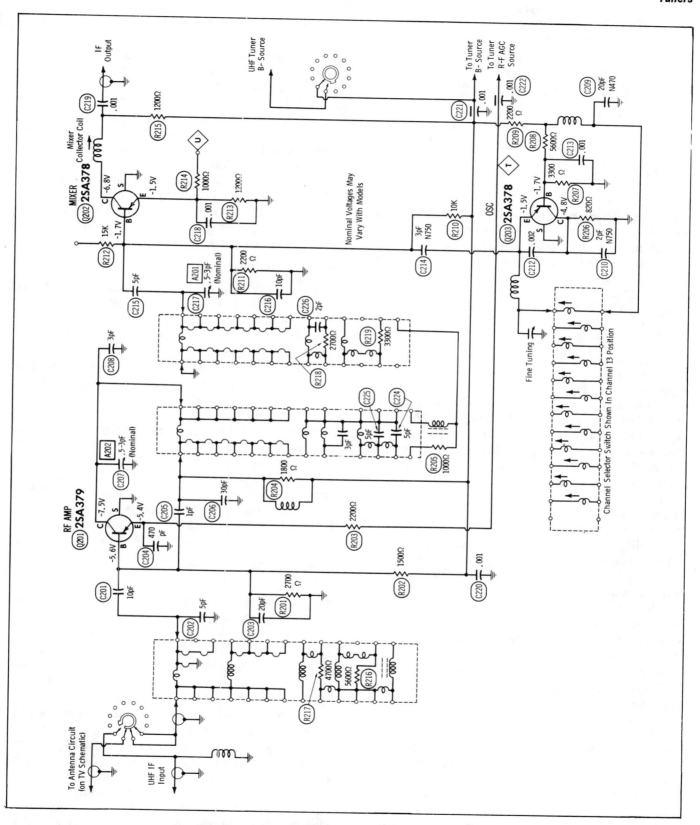

Fig. 12-6. Schematic of disc-type tuner shown in Fig. 12-5.

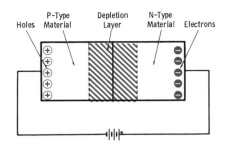

Fig. 12-7. Depletion area is formed when a diode is reverse-biased.

tube-type tuners still in use. We will limit our discussion of tube-type rf amplifiers to those circuits found in modern tube-type receivers. These circuits include the neutrode, tetrode, and nuvistor rf amplifiers. We will also discuss the transistor rf amplifier circuits used in the solid-state television receivers now being produced.

Neutrode RF Amplifiers

When a triode with a grounded cathode is used as an rf amplifier, the relatively high capacitance that exists between the plate and grid will tend to cause the stage to oscillate. One way this oscillation can be prevented is to neutralize the stage by providing a second feedback path. An out-of-phase signal can be applied through this path to the grid of the triode in order to cancel out the plate-to-grid feedback signal.

A newer version of the triode rf amplifier, called the neutrode circuit, is shown in Fig. 12-9. As the name implies, this circuit is a neutralized triode rf amplifier. Many of the tube-type tuners still being used employ the neutrode rf amplifier circuit.

The rf amplifier is neutralized by a trimmer capacitor connected from the low side of the plate coil to the control grid. The required 180° phase inversion is obtained through the plate coil. The amount of out-of-phase feedback is controlled by the adjustment of the neutralizing capacitor C_n. The neutralizing capacitor is adjusted to obtain the exact amount of feedback signal required for the most effective neutralization.

The performance of the neutrode compares favorably with that of cascode tuners. Field tests have shown that about 32 dB of gain can be achieved with less than 8 dB of noise.

Tetrode RF Amplifiers

A further development in the search for simpler circuits was the tetrode rf amplifier circuit shown in Fig. 12-10. This circuit was designed around a series of tetrodes which includes the 2CY5, 6CY5, etc. These tetrodes feature a high transconductance (8000 micromhos) with a noise figure approaching that of the rf triodes.

Circuitwise, the tetrode rf amplifier is almost identical to a pentode rf amplifier. However, there is no suppressor grid in the tube, and there is a small inductance in series with the screen-grid bypass capacitor. This inductance, which has a stabilizing effect on the stage, is actually in the lead of the capacitor itself. The tetrode tuner will provide about 35 dB of signal gain at a 6- to 8-dB noise figure, which is comparable to the performance of cascode rf amplifiers.

Nuvistor RF Amplifiers

The nuvistor is a high-mu triode intended for use as a grounded-cathode, neutralized rf amplifier. The nuvistor provides excellent performance in fringe areas where the signal level is very weak. The nuvistor triode provides exceptional signal-power gain and a very low noise factor. Also, the semiremote cutoff characteristics of the nuvistor reduces cross-modulation distortion from other vhf television signals.

An rf amplifier circuit employing a nuvistor is shown in Fig. 12-11. This circuit is very similar to the neutrode rf amplifier shown in Fig. 12-10. Trimmer capacitor C211 is the neutralizing capacitor. This circuit is found in many later tube-type receivers. The nuvistor is a ceramic tube housed in a metal shell and is much smaller than a conventional 7-pin miniature tube.

Transistor RF Amplifiers

A typical transistor rf amplifier is shown in Fig. 12-12. This circuit is similar to a grounded-cathode, triode rf amplifier. As in the case of the triode rf amplifiers, the transistor circuits usually require neutralization. Trimmer capacitor C214 is the neutralizing capacitor which provides out-of-phase feedback from the collector to the base of Q1. This feedback is used to cancel the in-phase feedback that is coupled from the collector to the base of the transistor due to junction capacitance.

The agc voltage applied to the base of the rf amplifier transistor in this circuit is known as forward agc. That is, as the signal strength increases, the agc voltage becomes more positive which causes the conduction of Q1 to increase. Due to the resistance in the collector circuit, the increase in collector current will cause the collector voltage to drop. This reduction in collector voltage causes the transistor to operate in its saturation region and, therefore, reduces the gain of the stage.

Many solid-state tuners now being used employ a field-effect transistor as the rf amplifier. The field-effect transistor, with its high input impedance and low noise-level characteristics, is ideally suited as an rf amplifier. The field-effect transistor used in the circuit shown in Fig. 12-13 is a dual-gate MOSFET. In this circuit, the dual-gate MOSFET may be considered as two separate transistors within a common case and arranged in a "cascode" configuration as illustrated in Fig. 12-14.

The rf signal is applied to the gate of Q1, which is a common-source amplifier. Transistor Q1 directly drives transistor Q2, which is connected in the common-gate configuration. The output is taken from the drain of Q2. This circuit has many of the characteristics and advantages of two tubes connected in the cascode configuration. Since transistor Q1 is loaded by the relatively low input impedance of Q2, it is not necessary to neutralize this rf amplifier.

The agc voltage for the rf amplifier is applied to gate G2. As is the case for tube-type rf amplifiers, a reverse-agc voltage is used with this circuit. As signal strength increases, the agc voltage becomes less positive and the drain current in both transistors is reduced. As a result of the lower drain current, there is a reduction in the gain of the rf amplifier.

OSCILLATOR CIRCUITS

The local oscillator used in a television tuner performs the same function as the local oscillator used in a superheterodyne radio receiver; it produces a signal which is beat against the incoming rf carrier to develop the if signal. However, the oscillators used in television tuners must have a high degree of stability and yet be able to oscillate over a wide range of frequencies in the vhf bands. The local oscillators in modern television receivers operate at a frequency 45.75 MHz above the video carrier of the channel being received. Since the oscillators used in television tuners operate at rather high frequencies with minimum loading, they require very little feedback to sustain oscilla-

tion. For this reason, the oscillator circuits found in most television tuners are relatively simple.

The oscillator in the tuner schematic shown in Fig. 12-4 is typical of the oscillators found in modern tube-type receivers. Capacitor C214, which is part of the tank circuit, provides feedback to sustain oscillation. The rest of the tank circuit is comprised of C212, C213, and the oscillator coil for each channel strip. Capacitor C212 also provides dc blocking for the grid of the oscillator tube, while capacitor C213 also functions as the overall oscillator adjustment.

The transistor oscillator circuit shown in Fig. 12-15 is typical of the oscillators found in solid-state tuners. This circuit is basically an ultra-audion Colpitts oscillator and is used extensively in solid-state tuners. Oscillation is sustained by feedback from the capacitive voltage divider network comprised of the internal collector-to-emitter capacitance (C_{CE}) and base-to-emitter capacitance (C_{BE}) of the transistor. Capacitor C212 shunts the internal capacitance of the transistor so that any changes in the internal capacitance will have minimal effect on the tuned circuits.

MIXER CIRCUITS

The function of the mixer circuit in a television tuner is to combine the incoming rf signal with the oscillator signal in order to produce the intermediate frequency (if) signal. To accomplish this heterodyning, the mixer stage must be a nonlinear amplifier. For this reason, the mixer stage is often referred to as the first detector. In tube-type tuners, the mixer stage usually employs a pentode or a tetrode. Even though these tube types have generally higher noise than triodes, this disadvantage is more than offset by their higher gain and stability characteristics.

In tube-type superheterodyne radio receivers, the functions of the oscillator and the mixer are usually performed by a single tube known as a pentagrid converter. The pentagrid converter is not used in television tuners due to the poor stability of the oscillator at television frequencies. In order to avoid excessive drift, converter tubes with separate oscillator and mixer sections are used in television tuners.

The mixer circuit shown in Fig. 12-4 is typical of the mixer found in tube-type tuners. The rf signal is inductively coupled to the grid of the mixer through a set of individual coils for each

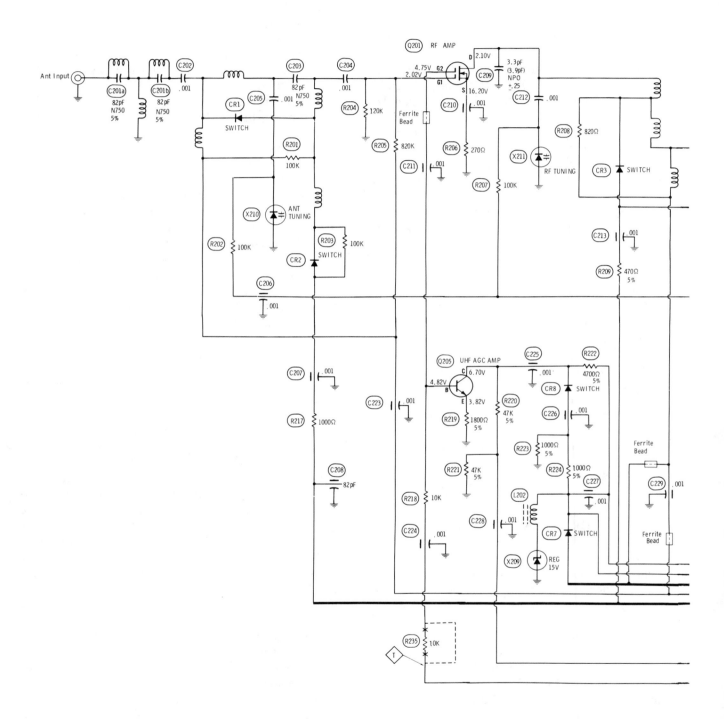

Fig. 12-8. Schematic of

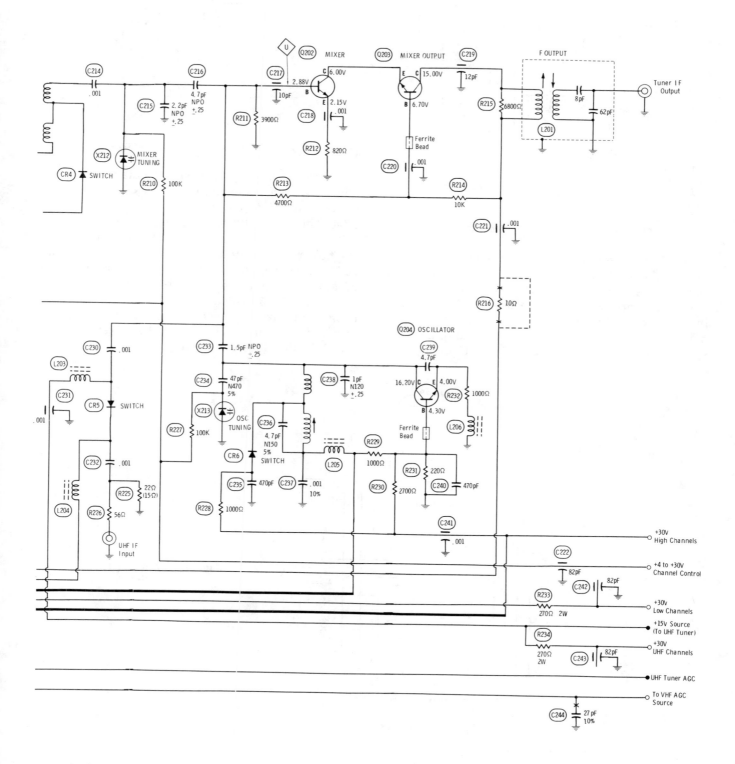

a varactor tuner.

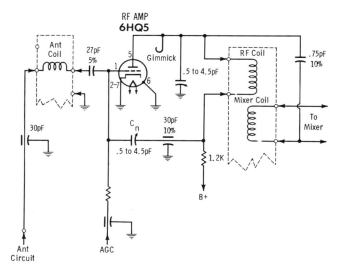

Fig. 12-9. A neutrode rf amplifier circuit.

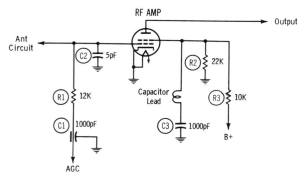

Fig. 12-10. The tetrode rf amplifier circuit.

channel. The oscillator signal is capacitively coupled to the grid by capacitor C210. The output of the mixer stage is coupled to the if strip by a pi-network comprised of the mixer plate coil and the inherent capacitance of the circuit. The mixer plate coil is usually adjusted for maximum response toward the high end of the if bandpass.

A typical solid-state mixer circuit is shown in Fig. 12-7. The operation of this circuit is very similar to the tube-type circuit depicted in Fig. 12-4. The signal from the collector of the rf amplifier is inductively coupled to the base circuit of the mixer. This coupling is provided by individual coils for each channel in the collector circuit of the rf amplifier and the base circuit of the mixer. Capacitor C215 provides dc isolation for the base of mixer Q202. It also couples the signal from the mixer input coil to the base of the mixer. Capacitor C219 provides dc blocking between the collec-

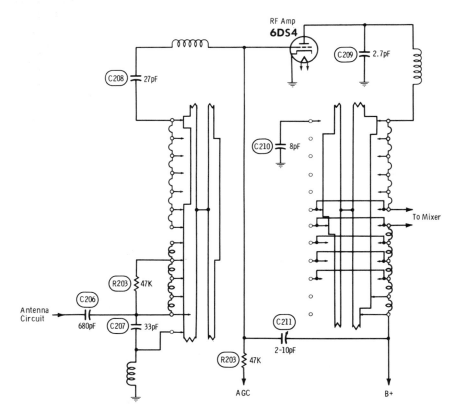

Fig. 12-11. A nuvistor rf amplifier circuit.

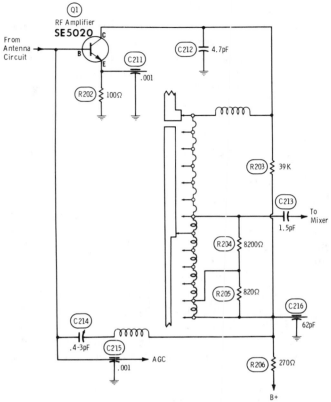

Fig. 12-12. Typical transistor rf amplifier circuit.

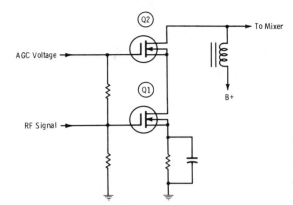

Fig. 12-14. Equivalent "cascode" configuration of a dual-gate MOSFET.

tor circuit of the mixer and the input of the if strip.

AUTOMATIC FINE TUNING

Many late-model television receivers, especially color sets, employ automatic fine tuning (aft) to adjust the local oscillator for optimum tuning. This is usually accomplished by connecting a varactor diode in the oscillator circuit as shown in Fig. 12-16. The function of the varactor diode in this circuit is very similar to its function in the

varactor tuner discussed previously. An aft control voltage is applied to the varactor diode in order to adjust the frequency of the oscillator. The control voltage changes the capacitance of the diode by increasing or decreasing the width of its depletion area.

The control voltage is developed by a discriminator circuit which reacts to a change in the 45.75-MHz video if frequency. Depending on the direction of the frequency change, the correction voltage developed by the discriminator will either increase or decrease. If the frequency of the oscil-

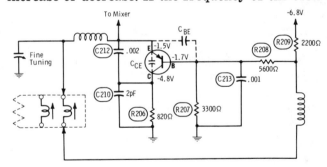

Fig. 12-15. Typical oscillator circuit found in a solid-state tuner.

Fig. 12-13. An rf amplifier employing a MOSFET.

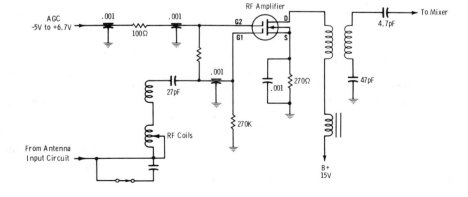

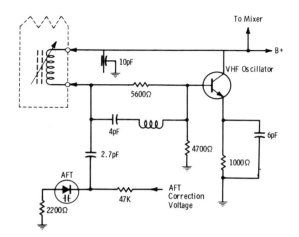

Fig. 12-16. A vhf oscillator employing a varactor diode for automatic fine-tuning control.

lator circuit shown in Fig. 12-16 should decrease, a more positive aft control voltage will be applied to the varactor diode. This will cause the capacitance of the diode junction to decrease, since its depletion area will be widened by the additional reverse-bias voltage. When the capacitance of the tuned circuit decreases, the frequency of the oscillator will increase to the point of optimum tuning. The range of the aft control voltage for a typical television tuner employing aft is from +1 volt to +8 volts.

An aft defeat switch is used to short out the aft control voltage for manual operation of the fine-tuning control. When the oscillator is operating outside the normal pull-in range of the aft circuit (50 to 100 kHz), it is usually necessary to defeat the aft voltage and manually fine-tune the oscillator to bring it within normal operating range.

UHF TUNERS

The tuners discussed so far have all been designed to receive the 12 vhf channels. Since the release of the 70 uhf channels in 1952, tuners have been produced to receive these channels. In fact, *all* receivers built since 1964 have been equipped with uhf tuners.

The reception of frequencies between 470 and 890 MHz presented many problems to the design engineer. The tubes available in 1952 were very inefficient at these frequencies. Also, the frequencies were too high for conventional lumped coil and capacitor tuning methods and too low for the waveguides and resonators employed in equipment

operating at 1000 MHz and beyond. The methods which have been developed tend to strike a compromise between the lumped inductance and capacitance of low-frequency circuits and the distributed inductance and capacitance of high-frequency circuits. It is well to remember that a straight length of wire can possess inductance and that capacitance can exist wherever there is a difference of potential between two surfaces.

When the uhf channels were first released, there were many vhf-only receivers already in the field. To enable these receivers to pick up uhf broadcasts, uhf converters were designed. A uhf converter is simply a uhf tuner and mixer-oscillator circuit in its own cabinet and with an output signal on one of the vhf channel frequencies. The output frequency was usually adjustable so it could be placed on one of the unused vhf channels.

Early receivers having turret tuners could be internally modified for uhf reception by removing unused channel strips from the turret and installing uhf channel strips in their places. However, receivers without turret tuners had to use the uhf converters.

Many methods of tuning have been used for uhf reception. One of the early tuning methods was the shorted quarter-wave or half-wave transmission line with a movable short. A transmission line possesses both inductance and capacitance, and these are distributed along the line. Any change in the length of the line will change the frequency to which it is tuned. To change the frequency, the movable shorting bar is ganged with the tuning control and dial indicator.

Another method of tuning, which is now used almost exclusively in modern uhf tuners, is a transmission line with a variable capacitor across the open end of the line. The method is as good as the sliding shorting-bar method; an added feature is that no moving contacts are needed.

Figs. 12-17 and 12-18 show a typical uhf tuner using capacitive tuning. Notice that a cavity-type arrangement is required due to the high frequencies involved. Nearly all recent uhf tuners employ a transistor oscillator and a diode mixer as shown in Fig. 12-18. An rf amplifier is not used in the uhf tuner since it is difficult to obtain efficient amplification at the uhf frequencies. The extra gain required when the uhf tuner is being used is provided by the rf-amplifier and mixer stages of the vhf tuner. The output of the uhf tuner is fed to the vhf tuner at the if frequency of the vhf tuner.

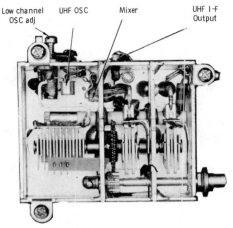

Fig. 12-17. Typical uhf tuner using capacitance tuning.

Therefore, the local oscillator in the vhf tuner is disabled when the uhf tuner is being used. In the uhf position, the inductances switched into the rf amplifier and mixer of the vhf tuner permit these stages to operate as if amplifiers.

QUESTIONS

1. Name the stages in a vhf tuner.

2. Most of the selectivity of the tuner is governed by what stage?

3. In the neutrode rf amplifier, how is feedback accomplished?

4. What circuit components are mounted on the strips of a turret tuner?

5. The principles of what two types of tuners are combined in the disc-type tuner?

6. How is tuning accomplished in a varactor tuner?

7. What is the purpose of an aft circuit?

8. How is forward agc employed in a transistor rf amplifier?

9. Why is there no rf amplifier stage in a uhf tuner?

Fig. 12-18. Schematic of the uhf tuner shown in Fig. 12-17.

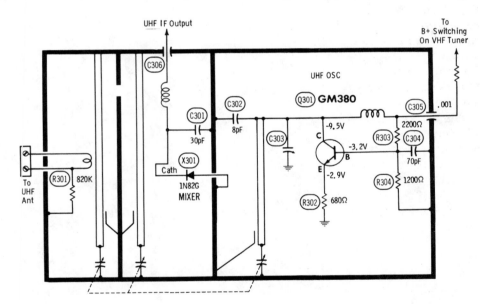

CHAPTER 13

Video IF Amplifiers and Detectors

INTRODUCTION

As discussed in Chapter 9, the composite television signal requires a channel 6-MHz wide. Like all superheterodyne receivers, the signal amplification and selectivity for a television receiver are obtained mostly in the intermediate-frequency (if) amplifiers. However, the television if amplifier system differs in one major respect from the familiar broadcast or short-wave receiver. The signal at the grid of the converter consists of two carriers which are separated by 4.5 MHz. These two carriers are the audio-modulated video carrier and the frequency-modulated sound carrier. The information contained in the modulation of these carriers must be separated and converted into a video voltage at the input of the video amplifier and an audio voltage at the input of the audio amplifier, respectively.

The video if amplifier system used in present-day television receivers processes both the video signal and the sound signal. This type of video if system, known as intercarrier if, is made possible by the fact that the video carrier and the audio carrier are always separated by a fixed difference of 4.5 MHz. The block diagram of a modern television receiver employing an intercarrier video if system is shown in Fig. 13-1. The response of the video if amplifier is made wide enough to accept both the video and the audio if carriers. The output of the second detector, the video detector, includes the demodulated video signal and a new 4.5-MHz if signal, which is frequency modulated in accordance with the audio signal. Since the video-amplifier response is wide enough, this frequency-modulated 4.5-MHz signal appears at the output of the video amplifier. A 4.5-MHz trap circuit prevents the audio signal from modulating the picture tube. The 4.5-MHz signal at the output of the video amplifier is fed to an fm detector, where it is demodulated and passed on to an audio amplifier. In color-television receivers, a separate sound if detector is used ahead of the video detector. The sound carrier is removed from the video if signal by a trap circuit located between the sound-takeoff point and the video detector. If the 4.5-MHz sound signal were allowed to reach the video amplifier, it would beat with the 3.58-MHz color signal and produce an objectionable 920-kHz pattern in the picture.

Requirements of the television if amplifier system are more complex than the if amplifiers in broadcast receivers. In a superheterodyne am receiver, the if amplifier must pass a band of frequencies only 10 kHz on either side of the center frequency. In fm receivers, the bandwidth need not be more than ±200 kHz. In a television receiver, however, the if amplifiers must pass a band approximately 5 MHz wide. In addition to this bandpass requirement, interfering signals from adjacent television channels must be rejected.

The major factors influencing the choice of an intermediate frequency for a television receiver are:

1. Bandwidth.
2. Selectivity.
3. Harmonics of the if frequency which might fall within the television band.
4. Direct if interference (externally generated signals having frequencies within the if band and passing through the rf amplifier).
5. Images due to fm or television stations on the image frequency.

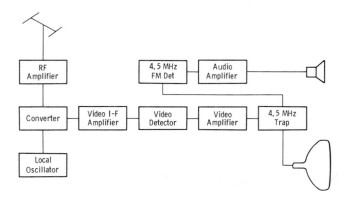

Fig. 13-1. Block diagram of an intercarrier video if system.

6. Cross modulation due to fm and television stations.
7. Oscillator radiation which will interfere with the operation of other television or fm receivers in the vicinity.

The choice of any particular intermediate frequency must be a compromise in view of the various factors just listed. However, when all of these factors are considered, the higher intermediate frequencies seem to be the better choice.

VIDEO IF SYSTEMS

The requirements of the video if system are:

1. The bandwidth must accept a total band approximately 5 MHz wide.
2. Frequencies beyond the edges of the passband must be rejected or attenuated so that they cannot interfere with the picture. These frequencies are the associated or co-channel sound carrier, the sound channel of the adjacent lower channel, and the video carrier of the next higher adjacent channel.
3. The response characteristic of the video if amplifier must be so shaped that the lower frequencies corresponding to the double-sideband portion of the transmission are properly attenuated to prevent overemphasis.

Required Response Characteristics of the Overall Video IF System

An analysis of the transmitter output characteristic in Fig. 9-3A reveals that the amplitude is constant from approximately 0.75 MHz below the video carrier to 4 MHz above the video carrier. If such a carrier and its sidebands are impressed

on a linear detector, the double-sideband nature of the region 0.75 MHz on either side of the carrier would cause high output from the detector for modulating frequencies from 0 to 0.75 MHz, and about half output for frequencies higher than 0.75 MHz. Fig. 13-2 shows the output from such an ideal detector. The output over the region from 0 to 0.75 MHz is twice that of the region from 1.25 MHz to 4 MHz; the output drops linearly from 0.75 MHz to 1.25 MHz.

To compensate for this increased low-frequency output due to the lower sideband, the overall response curve of the receiver should follow the linear slope of curve B in Fig. 13-3. This curve passes the 50-percent response point at the video-carrier frequency. In actual television receivers, the if response curve is usually shaped like C in Fig. 13-3.

As long as the area under curve B to the left of line X-X is equal to the area above curve B to the right of line X-X, these increments of the video-detector output will add and will produce the desired curve D. The ideal response characteristic for the high-frequency end of the band is a sharp cutoff at 4 Hz, as indicated by curves A and B. Most television receivers have a more sloping cutoff, as indicated by curves C and D in Fig. 13-3.

Although the reduction of high-frequency response causes some loss of fine detail in a test pattern, the loss is not noticeable when a moving scene is being televised. For economy, the passband in a black-and-white receiver can be reduced to 3 MHz without noticeably degrading the picture. However, the passband in a color receiver must be at least 4.2 MHz in order to accommodate the color signal.

Our discussion of the video and sound carriers and the associated adjacent-channel frequencies, up to this point, has been concerned with their position in the transmitted rf signal. Since the local oscillator of the television receiver operates at a frequency higher than the incoming rf signal, the co-channel carrier frequencies and the adjacent-channel frequencies will be inverted in order when they appear as if frequencies at the output of the first detector.

The overall video if response curve for a modern television receiver is shown in Fig. 13-4. This curve shows the sound and video if frequencies of the desired channel and the position of the adjacent-channel if carriers which might interfere with the operation of the receiver. Trap circuits are used to attenuate the curve at the associated

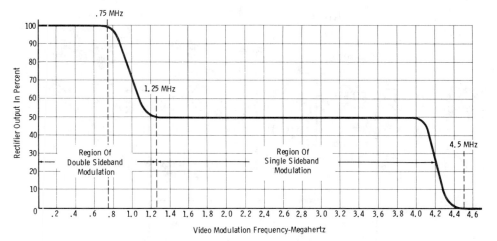

Fig. 13-2. Output of an ideal (linear) detector rectifying the vestigial-sideband, modulated video carrier.

sound if frequency and the adjacent-channel if frequencies.

The choice of an if frequency is limited by the television system and the frequency assignments. The frequency chosen must be between the highest video-frequency component on one end, and the lowest television-channel frequency assignment on the other. In other words, the if frequency cannot be lower than approximately 8 MHz or higher than 50 MHz. Although some early television receivers had video if frequencies as low as 15.2 MHz, almost all present-day receivers in the United States employ a standard video if frequency of 45.75 MHz and a sound if frequency of 41.25 MHz.

Methods of Obtaining Wideband Response in Video IF Systems

Many types of coupling networks can be used between the video if amplifier stages to accomplish the wideband response required. The two major methods of coupling found in receiver design are overcoupled transformers with shunt resistance loading and stagger-tuned circuits.

Video IF Amplifiers With Overcoupled Transformers—If the primary and secondary of an interstage transformer are tuned to the same frequency, and the coupling between the circuits is progressively increased, a double-humped resonance curve will occur after the critical coupling point has been passed. These double humps can be smoothed to a flat-topped response curve by loading the primary and secondary circuits with the correct resistance. This results in lower Q and, consequently, lower gain than would be obtained with a narrower passband. Fig. 13-5 shows an example of an if amplifier system using overcoupled transformers.

The amplifier in Fig. 13-5 uses three transformer-coupled stages. Transformers L2 and L3 are overcoupled and then loaded with resistance in the secondary circuits to produce the flat-topped bandpass characteristic. No primary shunt

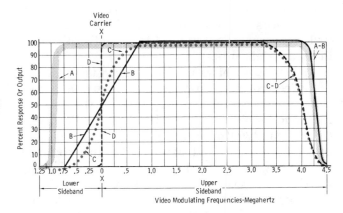

Fig. 13-3. Overall receiver-response characteristic required to compensate for vestigial-sideband modulation.

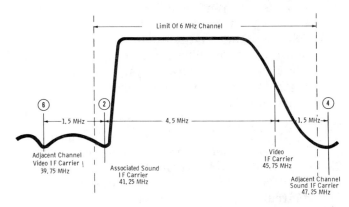

Fig. 13-4. Overall video if response curve showing location of if trap frequencies.

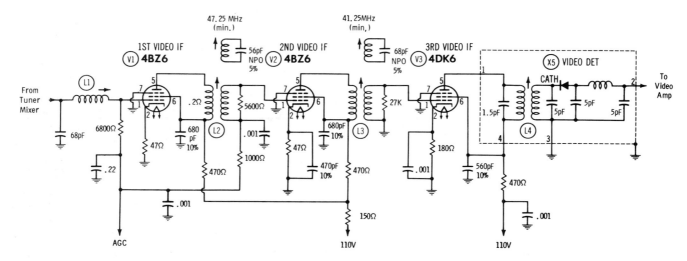

Fig. 13-5. Tube-type video if amplifier system employing overcoupled transformers.

resistance is shown because the plate-to-cathode resistance of the tubes provides the required primary loading. The transformers used in this circuit are bifilar transformers, which are constructed by simultaneously winding the primary and secondary coils side by side. Since the coils occupy essentially the same space on the coil form, both the primary and the secondary can be tuned by the same slug. This makes the if stages somewhat less difficult to align.

Decoupling, or isolation, of each stage from all the others, except for the signal path, is important in television if amplifiers. If the second video if amplifier tube (V2 in Fig. 13-5) were to receive a portion of the output signal from the third video if amplifier (V3), the entire if strip would oscillate. The feedback signal could be returned through the B+ supply line, through the filament supply line, or through capacitive coupling between components in the different stages. Since no signal amplification is possible when the amplifiers are oscillating, adequate decoupling must be provided in the system. The B+ line in Fig. 13-5 is, therefore, decoupled at each stage in the if amplifier strip. A 470-ohm series resistor and a bypass capacitor isolate the individual stages.

Video IF Amplifiers With Stagger-Tuned Circuits—Another method of designing a video amplifier to meet wideband requirements is to couple the stages by tuned circuits which are tuned to different frequencies within the passband. If the tuned circuits between stages are all tuned to the same frequency, the bandwidth decreases as more stages are added, and the overall response grows

more peaked or selective. However, if the individual circuits are stagger-tuned about the center frequency and the Q values of the tuned circuits are properly adjusted, the desired bandwidth can be obtained and a satisfactory overall curve will result.

The curves in Figs. 13-6B and 13-6C illustrate the effects of the tuning frequency on the bandwidth of a pair of identical circuits. Coils L1 and L2 of the circuit shown in Fig. 13-6A are first tuned to the same frequency and then to frequencies separated by half the bandwidth of the individual circuits. The bandwidth of a circuit is defined as the frequency spread between the points on the resonance curve at which the response is 0.7 of that at resonance. When both circuits are tuned to the same frequency (Fig. 13-6B), the overall bandwidth of the amplifier will be 64% of the bandwidth of the individual circuits. When the two circuits are stagger-tuned, the overall bandwidth increases to 140% of the bandwidth of the individual circuits.

Although we have used two circuits to explain the effects of stagger-tuning, a particular television video if amplifier may involve the staggering of more circuits to achieve the required 4- to 6-MHz bandwidth. In the two circuits just illustrated, the individual bandwidths and circuit Q values have been made identical for simplicity. In television receiver circuits, however, the Q values are varied to produce resonance curves and stage gains which will, in turn, cascade to produce the overall curve in Fig. 13-4. Fig. 13-7 shows a typical stagger-tuned video if circuit. Notice the

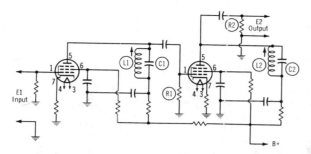

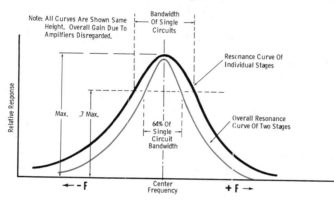

(A) Single-tuned coupling circuits between tube-type video if stages.

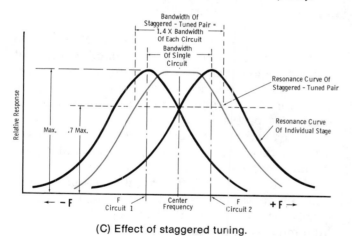

(B) Effect of cascading circuits on the same frequency.

(C) Effect of staggered tuning.

Fig. 13-6. The effect of tuned circuits on bandwidth.

different frequencies to which L1 through L7 are tuned. Three stages of if amplification are needed in this circuit to obtain sufficient if gain and the necessary bandwidth characteristic.

While most modern tube-type television receivers employ either two or three if stages, some solid-state sets may have as many as four. A stagger-tuned video if amplifier used in a transistor television receiver is shown in Fig. 13-8. The required bandpass is obtained in this receiver by

using four if stages tuned to frequencies ranging from 42.8 MHz to 45.3 MHz. Although this particular solid-state receiver employs four video if stages, many transistor models use only three stages. A two-stage if amplifier used in a tube-type color-television receiver is shown in Fig. 13-9.

In some late-model solid-state television receivers the video if stages are included in one or more integrated circuits (ICs). Fig. 13-10 shows the video if amplifiers, video detector, video amplifier, and agc functions contained within two ICs. The if output from the tuner is fed through an if trap coil to the input of the 1st and 2nd video if amplifier stages. These first two video if amplifier stages are part of IC HA1144 which also includes the agc functions. The output from the 2nd video if amplifier is coupled to the 3rd video if amplifier by single-tuned transformer T201. The 3rd video if amplifier stage is part of IC HA1167 that includes the video detector and video amplifier. The output of the 3rd video if amplifier is tuned by coil L207 and coupled to the video detector through capacitor C215. The only tuning adjustments for the complete video if system are provided by transformer T201 and coil L207.

Rejection of Undesired Adjacent-Channel and Co-Channel Carriers

In our discussion of the required overall response curve for the video if amplifier (Fig. 13-4), we learned that the response must be reduced at the position which would be occupied by the sound carrier of the same channel being received, the video carrier of the next higher adjacent channel, and the sound carrier of the next lower adjacent channel. The low-response points or "notches" at the intermediate frequencies corresponding to these carriers can be produced in the video if amplifiers by one or a combination of the following circuits:

1. Series-tuned trap circuits.
2. Parallel-tuned circuit absorption traps.
3. Cathode-circuit or degenerative traps.
4. Bridged-T networks as coupling elements and rejection circuits.

The first two trap circuits function like the interference-elimination traps used in radio receivers to eliminate image interference or interference by strong local stations. The degenerative-trap principle is often used in audio circuits for compensation. The fourth type, the bridged-T trap,

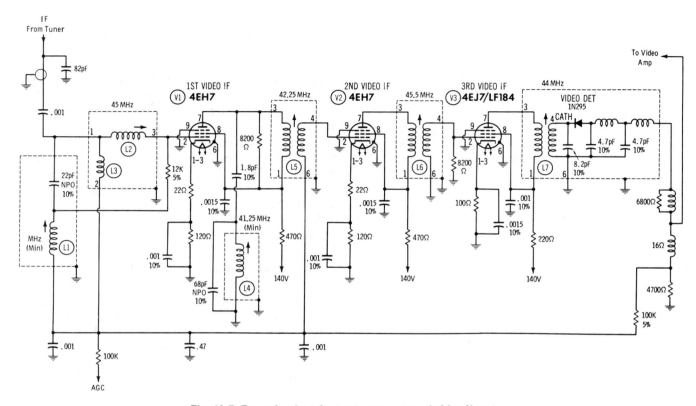

Fig. 13-7. Example of a tube-type stagger-tuned video if system.

is like a null bridge often used for measurement purposes.

Series-Tuned Traps—The acceptance-type series-tuned trap acts as a short circuit across the system at its resonant frequency. The impedance at the resonant frequency is very low and equals the ac resistance of the circuit components. In the if amplifier shown in Fig. 13-5, the circuit comprised of coil L1 and the input capacitance of tube V1 is used to adjust the slope at the high-frequency end of the if bandpass curve. In this manner, the point at which the video carrier drops to 50% of maximum response (for compensation of the vestigial-sideband effect) can be controlled. When this adjustment is made, the curve drops fast enough to reject the next-lower channel sound carrier.

Parallel-Tuned or Absorption-Type Traps—The absorption-type trap is the most frequently used rejection circuit in television if amplifiers. These traps consist of parallel-tuned circuits coupled to the video if tuned circuits. Absorption traps are usually wound on the same coil form as the video if transformer, but they are tuned by a separate slug. At the resonant frequency, a high circulat-

ing current is developed in the trap. Because the trap reacts through its inductive coupling, the load impedance and amplification of the stage are reduced at the resonant frequency of the trap. Transformers L2 and L3 in Fig. 13-5 include absorption-type traps.

Cathode-Circuit or Degenerative Traps—Another method of reducing amplifier gain at a particular frequency is by selective degeneration. This is done by placing a parallel-resonant combination in the cathode circuit of one of the if stages. At the resonant frequency of the trap, a high impedance appears in the cathode circuit and causes degeneration, which reduces the amplification of the stage to a very low value. In other words, the trap acts like a large unbypassed cathode resistor at the resonant frequency. The rejection figure of such a degenerative trap can never be greater than the gain of the stage.

Bridged-T Networks—Fig. 13-11 shows a network of circuit elements known as the bridged-T network. This circuit uses a T-shaped branch consisting of two capacitors and a resistor (C1, C2, and R1) bridged by inductor L2. This arrangement acts like a bridge circuit, but has the added

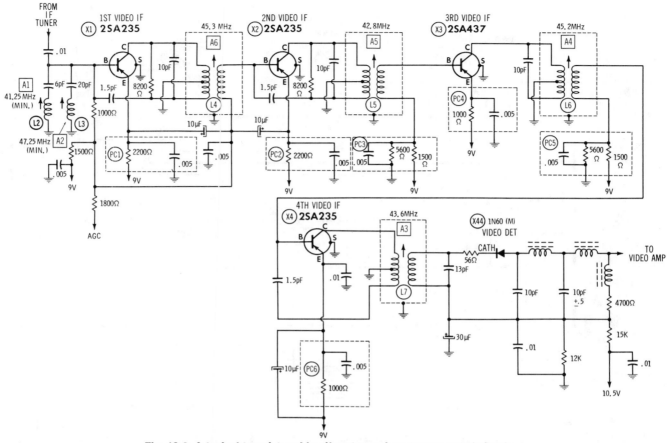

Fig. 13-8. A typical transistor video if system using stagger-tuned circuits.

advantage that the input and output circuits have a common terminal at ground potential.

Balance (for a null or low output) occurs when the reactance of the variable inductor is equal to the reactance of the capacitors in series, and when the resistance in the center leg of the T is approximately one-fourth of the parallel impedance of the tuned circuit. The parallel LC circuit acts

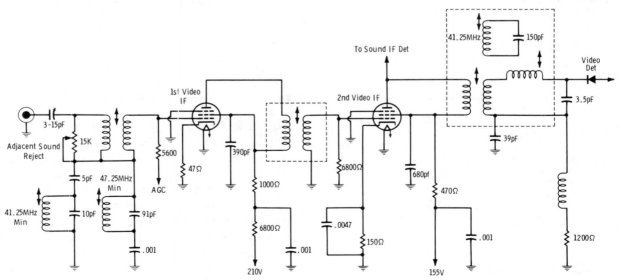

Fig. 13-9. A two-stage video if amplifier used in a tube-type color receiver.

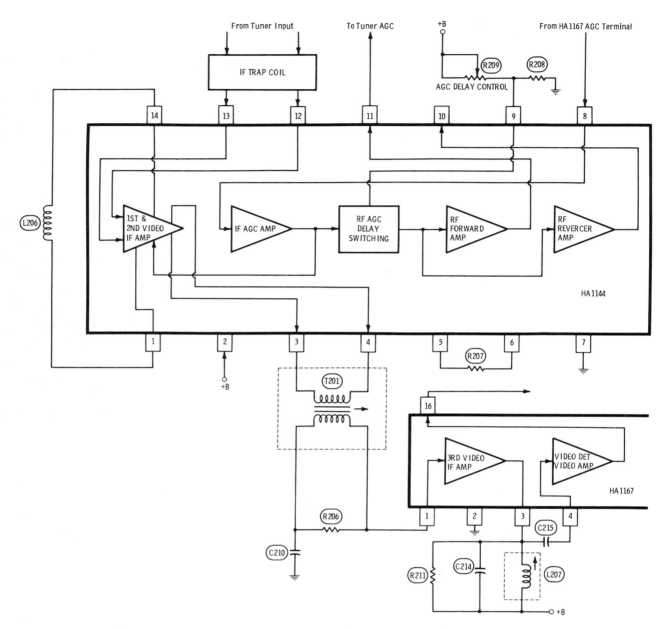

Fig. 13-10. A video if amplifier system employing ICs in a solid-state television receiver.

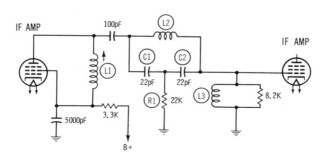

Fig. 13-11. The bridged-T network used as a trap in a tube-type if amplifier circuit.

as an antiresonant trap between the input and output terminals, while the resistance leg provides a secondary balance for the resistance losses of the circuit at frequencies other than the rejected frequency. Much sharper null notches and greater rejection can be obtained with the bridged-T connection than with the antiresonant trap alone. Another variation of the bridged-T circuit (not illustrated) employs a center-tapped coil for the inductance branch and a single trimmer for the capacitance branch.

VIDEO DETECTORS

The video if amplifier is followed by the video detector, which is essentially the same as the second detector employed in am broadcast or short-wave radio receivers. However, two significant circuit differences must be taken into consideration for the video detector: (1) a means of compensation must be used to prevent the loss of higher video frequencies, and (2) the polarity of the detector output must be considered.

Most early-model television receivers employed a vacuum-tube video detector and up to three stages of video amplification. Solid-state diodes have been used exclusively as video detectors for the past several years. Most modern, black-and-white, tube-type tv receivers employ only one stage of video amplification, although two or three video amplifiers are used in the majority of color sets. Solid-state television receivers commonly employ two to four video-amplifier stages.

Since each amplifier produces a 180° phase inversion in the applied signal, the number of video-amplifier stages following the video detector is important. Of course, if a particular stage is a cathode follower or an emitter follower, there is

no signal inversion for that stage. The polarity of the video signal applied to the picture tube must be correct; otherwise, the picture will be negative (black will appear white and vice versa). The polarity of the signal at the picture tube is determined by the polarity of the video-detector output and the number of amplifier stages which provide signal inversion. The video signal may be applied either to the grid or to the cathode of the picture tube, depending on the signal polarity. In other words, if the signal has the wrong polarity for application to the grid, it can be applied to the cathode in order to produce the correct picture phase. Fig. 13-12 shows an example of a video detector and amplifier system which applies the video signal to the cathode of the picture tube. The cathode is used as the driven element in practically all receivers currently being produced.

If there are an even number of video amplifier stages providing phase inversion, and the signal is applied to the cathode of the picture tube, the output of the video detector must be positive going. In other words, when a darker scene is being televised, a more-positive signal is applied to the cathode of the picture tube, causing the screen to become darker. If an odd number of phase-invert-

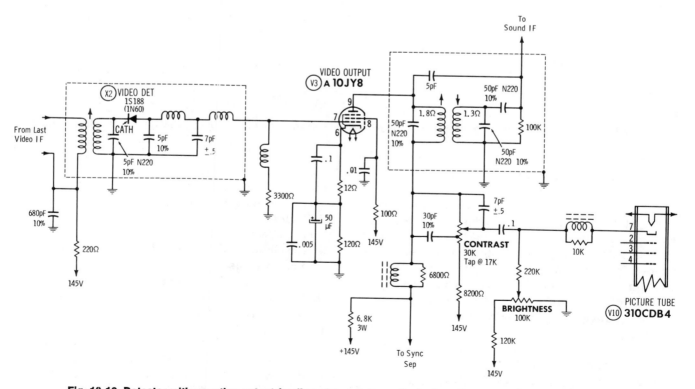

Fig. 13-12. Detector with negative output feeding picture-tube cathode through a single video-amplifier stage.

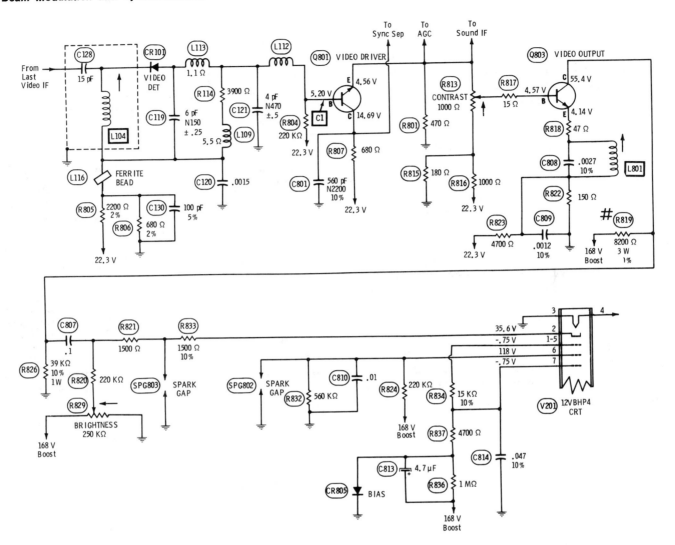

Fig. 13-13. Solid-state video-amplifier system with single phase-invering stage requires negative output from video detector.

ing video-amplifier stages are used to apply the video signal to the cathode of the picture tube, the video-detector output must be negative.

Fig. 13-13 shows the video detector and amplifier system used in a late-model solid-state television receiver. Since the video-driver stage is an emitter follower, there is no phase inversion at this stage. However, the video-output stage does provide phase inversion of the signal applied to the cathode of the picture tube. Because there is only one phase-inverting video-amplifier stage, the output from the video detector must be negative.

A significant difference between an am detector used in a radio receiver and a video detector is in the value of the load resistor. In a typical am broadcast receiver, the detector-diode load ranges from 0.5 to 2 megohms and maintains this high load resistance over the range of frequencies required for sound reproduction. In most instances, frequencies no higher than 5000 Hz are involved. The video-detector circuit, however, must provide a flat response to at least 4 MHz. The capacitance of the detector diode and its associated circuitry prevent the use of a high diode-load resistance. At these high frequencies, the reactance of the detector circuit would become lower than the load resistance and thus bypass the high frequencies to ground.

In order to maintain a flat response at the high frequencies, the value of the load resistance is made low, and compensating elements are used to produce a resonant rise of circuit impedance at

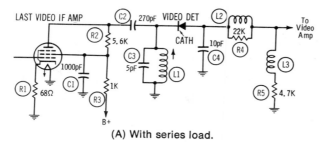

(A) With series load.

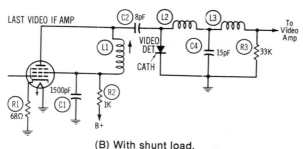

(B) With shunt load.

Fig. 13-14. Basic video-detector circuits with high-frequency compensation.

the high end of the video band. Fig. 13-14 shows two basic video-detector circuits used in modern television receivers. The load for the circuit shown in Fig. 13-14A includes resistor R5 and the high-frequency compensation network consisting of C4, L2, L3, and R4 connected in series with the diode. The shunt-connected circuit in Fig. 13-14B is loaded by resistor R3 and the compensation network, consisting of L2, L3, and C4. The inductors used in these compensation networks are usually referred to as peaking coils. In Fig. 13-14A, L2 is used as a series-peaking coil, and L3 is a shunt-peaking coil. Peaking coils are employed for frequency compensation in each video stage. A more

complete discussion of their function will be presented when video amplification is considered in Chapter 15.

QUESTIONS

1. When the high frequencies of the television signal are lost, what happens in the reproduced picture?

2. What two coupling methods are used to obtain the wide-band response required from the video if amplifiers?

3. What are the results when the primary and secondary of a coupling transformer are loaded with a shunt resistance?

4. What is meant by staggered tuning?

5. Why are traps necessary in the video if circuits?

6. When the video signal is applied to the cathode of the picture tube, and there are two common-emitter video-amplifier stages, what must be the polarity of the video-detector output?

7. What must the polarity of the video signal be if it is applied to the grid of the picture tube? The cathode?

8. Why is decoupling important in the video if system?

9. What is meant by an absorption-type trap?

10. Why must a low-impedance load be used with video detectors?

EXERCISE

1. Draw the overall if response curve and show:
    (a) The frequency limits.
    (b) The location of the video carrier frequency.
    (c) The location of the trap frequencies.

CHAPTER 14

Sound IF Amplifiers and Audio Detectors

The sound if system of a television receiver is similar to the if system of an fm radio receiver. However, two major differences from the standard fm radio receiver will be noted.

1. The intermediate frequency for fm receivers has been standardized at 10.7 MHz, whereas the sound if frequency for television receivers has been standardized at 4.5 MHz. This is the frequency difference between the video and sound carriers and is held constant at the transmitter. The sound if frequency will always be 4.5 MHz lower than the video if frequency.
2. The deviation of the television sound carrier for maximum modulation has been established at 25 kHz (a total sweep of 50 kHz), whereas the maximum standard deviation for standard fm broadcasting has been set at 75 kHz (a maximum sweep of 150 kHz). Because a lower deviation is employed for television sound, a narrower passband can be used for the television sound if system, and a shorter linear region can be used for the detector. The passband of a typical sound if amplifier is about 150 kHz, and the linear region of the detector usually does not exceed 100 kHz. A high-quality fm receiver might have an if passband of 300 kHz and a linear range in the detector of as much as 2 MHz.

SOUND IF TAKEOFF

The sound takeoff point for intercarrier sound systems usually follows the video detector, since the sound detector requires a definite amount of signal to provide a noise-free output. The amplification for this signal is provided by the sound if amplifier, the video amplifier, or a combination of both.

The sound takeoff point in most modern receivers is either at the video-detector output (Fig. 14-1A) or following the video output stage (Fig. 14-1B). The 4.5-MHz signal is coupled from the resonant takeoff point to the sound if amplifier. When the sound takeoff point is in the plate circuit of the video amplifier, there is the advantage of the additional gain supplied by the video amplifier.

As discussed in Chapter 13, the sound takeoff point in a color-television receiver must precede the video detector. This permits the use of a 4.5-MHz sound trap ahead of the video detector and prevents the 4.5-MHz sound signal from beating with the 3.58-MHz chroma signal at the video detector. The 920-kHz difference signal produced by beating these two frequencies would produce an objectionable pattern on the screen of the picture tube. Since the sound takeoff point precedes the video detector, it is necessary to use a separate detector to develop the 4.5-MHz sound if signal. Fig. 13-9 shows the sound takeoff point following the second video if amplifier.

TYPICAL SOUND IF SYSTEMS

The function of the sound if amplifier circuits is to provide sufficient amplification of the 4.5-MHz sound if signal before it is applied to the audio detector. While early television receivers employed as many as three sound if stages, most modern tube-type sets have only one sound if stage. Transistor sound if systems are usually comprised of two or three stages.

Typical sound if circuits are shown in Figs. 14-2 and 14-3. The 4.5-MHz signal obtained from the resonant circuit at the sound takeoff point (Fig.

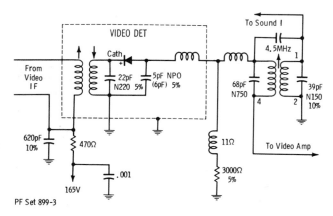

(A) Following the video detector.

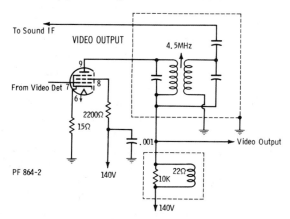

(B) Following the video-output stage.

Fig. 14-1. Typical sound-takeoff points.

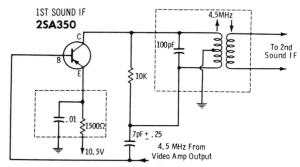

Fig. 14-3. Typical transistor sound if circuit.

14-1) is amplified by the sound if circuit and developed across the 4.5-MHz if transformer. The tube-type circuit shown in Fig. 14-2 is a single-stage if amplifier and its output is applied directly to the audio detector. The transistor circuit shown in Fig. 14-3 is the first stage of a two-stage sound if system. The signal developed across the sec-

ondary winding of the 4.5-MHz transformer is applied to the second sound if stage which is similar to the first stage. The 7-pF capacitor is used to neutralize the stage and prevent it from oscillating.

Another type of sound if amplifier is shown in Fig. 14-4. In this circuit, the triode section of the triode-pentode tube functions as the sound if amplifier, while the pentode section serves as the video amplifier. Since the sound if amplifier is a triode operating at a relatively high frequency, it must be neutralized. Neutralization is accomplished by the tapped plate coil and the 4.5-pF capacitor.

A reflex amplifier, which acts as both the sound if amplifier and the audio amplifier, is shown in Fig. 14-5. In this unique circuit, the output signal from the audio detector is applied to the input of the sound if amplifier, so that the tube can amplify both signals simultaneously. The plate load for the 4.5-MHz signal is the transformer, and the plate load for the sound signal is the 27K resistor.

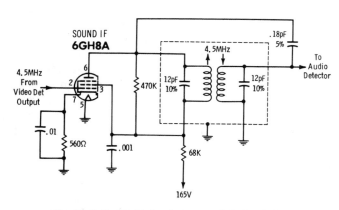

Fig. 14-2. Typical tube-type sound if circuit.

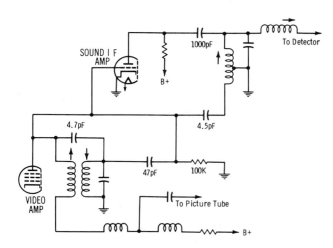

Fig. 14-4. A neutralized triode used as a sound if amplifier.

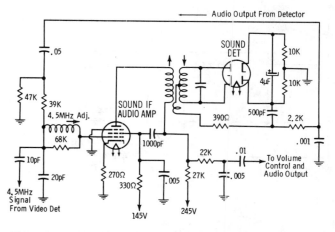

Fig. 14-5. A reflex amplifier which amplifies the sound if and audio signals.

Many late-model television receivers utilize integrated circuits in the audio section. In Fig. 14-6, a single integrated-circuit (IC) chip functions as the sound if amplifier, audio detector, and audio amplifier. This single integrated-circuit chip contains the equivalent of several transistors, diodes, and resistors. The tuning components, as well as many of the coupling and bypass capacitors, are external to the integrated-circuit unit. The sound if section is comprised of three differential-amplifier stages which provide amplification and limiting for the if signal. The audio detector section of the integrated circuit is a ratio detector. The detected audio signal is then amplified by the audio-amplifier section of the IC which is equivalent to a Darlington pair direct-coupled to an emitter-follower stage.

AUDIO DETECTORS

The television receivers presently in use employ at least four types of fm demodulation circuits: the ratio detector, the gated-beam detector, the locked-oscillator detector, and the delta sound system. The gated-beam and locked-oscillator detectors are often called quadrature detectors.

Ratio Detectors

The ratio detector was a popular type of fm detector in tube-type television receivers and is used extensively in solid-state receivers, especially those that employ discrete transistors. Besides the function of fm detection, the ratio detector features an inherent rejection of amplitude modulation.

The ratio detector circuit is shown in Fig. 14-7. The transformer, consisting of primary L2 and secondary L3, couples the output of the sound if amplifier to diodes X1 and X2. Both windings of the transformer are tuned to the center if frequency by separate slugs. The primary voltage is also coupled to the center tap of the secondary winding by capacitor C5. The reactance of C5 is negligible at the sound if frequency.

When the secondary winding is resonant at the incoming signal frequency, the primary voltage at the center tap of the secondary winding is in quadrature (90° out of phase) with the voltage across each half of the secondary. The phase relationship varies as the signal frequency changes. The resultant voltages due to changes in frequency are applied to diodes X1 and X2. The diodes are

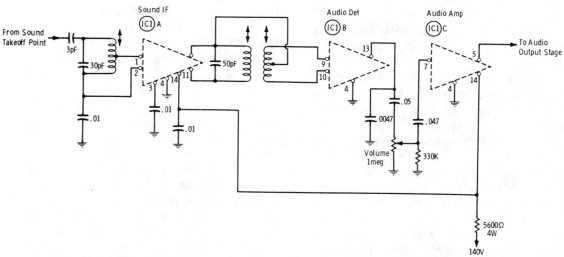

Fig. 14-6. Integrated circuit used as a sound if amplifier, audio detector, and audio amplifier.

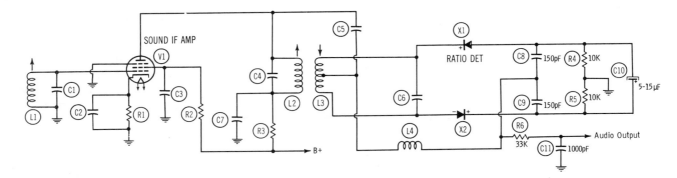

(A) The basic circuit.

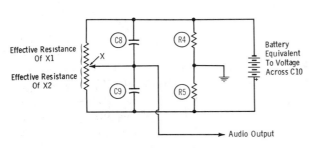

(B) Equivalent circuit of the ratio detector.

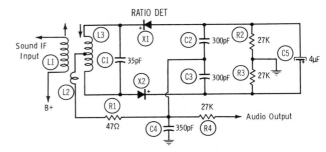

(C) Most common variation of the basic circuit.

Fig. 14-7. Ratio-detector circuits.

effectively connected in series across load resistors R4 and R5. An electrolytic capacitor (C10) is used across the load resistors.

The conduction of the diodes in series produces a stable charge across capacitor C10. It is convenient to think of stabilizing capacitor C10 as a form of bias battery across the two diodes in series (see Fig. 14-7B). The diodes may be considered as being variable resistances across this battery. Each resistance is determined by the resultant voltage impressed on that diode. With frequency modulation, the resultant voltage impressed on one diode increases as the voltage impressed on the other diode decreases, and vice versa. Therefore, the variable resistance of the diodes has been represented as a potentiometer in Fig. 14-7B.

When an unmodulated carrier is received, the voltage across the top half of the secondary is 90° out of phase with the voltage at the center tap of the secondary winding. The resultant of the two voltages is applied to diode X1. At the same time, the voltage across the bottom half of the secondary is 90° out of phase with the center tap voltage. The resultant voltage of these two voltages is applied to diode X2. Because the phase angles are equal, the voltages applied to the diodes are equal.

Therefore, the diodes conduct equally and their resistances are equal. The potential at their junction (X in Fig. 14-7B) does not change. Consequently, there is no audio output.

When the frequency of the incoming signal increases because of modulation, the phase difference between the primary voltage and the voltage across the top half of the secondary is less than 90°. At the same time, the phase difference between the primary voltage and the voltage across the bottom half of the secondary becomes more than 90°. This causes the voltage impressed on diode X1 to be higher than the voltage impressed on X2. As a result, the effective resistance of X1 decreases while the effective resistance of X2 increases. This has the same effect as if the arm of the potentiometer in Fig. 14-7B were moved upward. The output voltage swings in a negative direction.

When the frequency of the incoming signal decreases, the arm of the potentiometer is effectively moved downward and the output voltage swings in a positive direction. In this way, fm detection takes place. It can be seen that the ratio-detector output is proportional to the ratio between the amplitudes of the impressed voltages. Since the

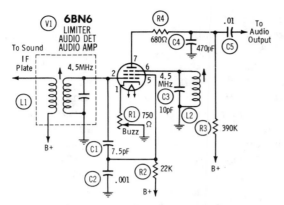

Fig. 14-8. The 6BN6 gated-beam detector.

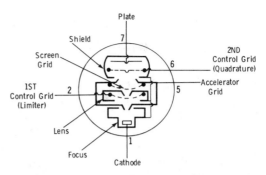

Fig. 14-9. Internal construction of the 6BN6 tube.

ratio of the voltages instead of the differences is used, amplitude modulation of the incoming signal has little effect on the output.

When the incoming signal is amplitude modulated, the voltages on the diodes change in the same direction. This is equivalent to an identical change in the effective resistance of *both* halves of the potentiometer in Fig. 14-7B. The arm does not move, and there is no output voltage.

Fig. 14-7C shows the most common variation of the basic type of ratio detector. The major difference is that a tertiary winding (L2), instead of a capacitor, supplies the quadrature voltage. Winding L2 is closely coupled to the primary winding and, therefore, the voltage phases on the two windings are substantially the same. The phase relationship of the coupled voltage to the voltage of the secondary is still 90° at the center frequency.

The Gated-Beam Detector

The circuit in Fig. 14-8 presents a unique method of fm detection. A special tube, the 6BN6, was designed to function as a limiter, audio de-

tector, and audio amplifier. The construction of the 6BN6 is shown in Fig. 14-9. Its cathode provides a thin beam of electrons that are guided and accelerated toward the plate. A limiter grid and a quadrature grid in the path of the electrons act as gates, and either grid can cut off the electrons. Both grids have a sharp cutoff feature, allowing a small signal voltage to drive the tube from cutoff to saturation. Amplitude modulation is rejected or limited because large signal variations have no greater effect on plate current than small signal variations.

As the plate current of the tube flows past the quadrature grid, a voltage is induced on this grid by space-charge coupling. The 4.5-MHz tuned circuit (L2 and C3) is connected to the quadrature grid. This circuit causes the alternating voltage on the quadrature grid to be 90° out of phase with the incoming if signal on the limiter grid when the signal is unmodulated.

When modulation is impressed on the if signal, the phase difference between the limiter and quadrature grids varies above and below 90°. When the input signal goes positive, the limiter grid allows current to pass. A short time later, depending on the instantaneous phase difference, the quadra-

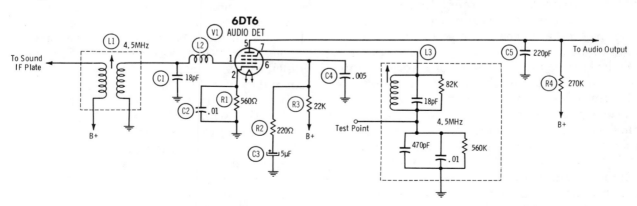

Fig. 14-10. The locked-oscillator detector.

ture grid goes positive and allows current to pass. When the input signal goes negative, the plate current stops. The average plate current will depend on the length of time the plate current flows on each positive cycle. When the audio modulation varies, the phase difference, the conduction time, and the average plate current will all vary proportionately. These plate-current variations contain the same modulation in the form of an amplified audio signal large enough to drive the output stage directly.

The gated-beam detector usually has an adjustable resistor in the cathode circuit. The resistor adjusts the tube bias so that the incoming signal will drive the limiter grid into both saturation and cutoff. This resistor is called the buzz control because it limits the buzzing sound caused by sync signals getting through to the audio section.

The gated-beam detector becomes very inefficient on weak signals. When the incoming signal at the limiter grid drops below a certain voltage, limiting stops, and sync buzz and noise are reproduced.

The Locked-Oscillator Detector

A circuit which operates much like the gated-beam detector is shown in Fig. 14-10. The similarity between the gated-beam and locked-oscillator circuits can be clearly seen by comparing Figs. 14-8 and 14-10. The most obvious difference is that the locked-oscillator lacks a buzz control. The locked-oscillator circuit employs a special sharp cutoff pentode, such as a 6DT6 or a 6HZ6.

The 6DT6 and 6HZ6 are not as complex as the 6BN6 and are much like an ordinary pentode, except that the control and suppressor grids can both sharply cut off the plate current. During the reception of moderate or strong signals, quadrature-grid detection takes place much the same as it does in the gated-beam detector, except for a few differences in limiting.

The locked-oscillator mode of operation of this circuit does not come into play until the input signal becomes fairly weak. Below a certain signal amplitude, the detector will break into 4.5-MHz oscillation. This oscillation tends to keep the signal amplitude constant in the detector, despite amplitude variations which may occur in the input signal because of noise or fading.

The circuit can oscillate because of positive feedback from the quadrature-grid circuit to the

control-grid circuit through interelectrode capacitance of the tube. A similar arrangement using a 6BN6 cannot be set up, because the interelectrode capacitance of this tube is too small to provide the required amount of feedback.

The input signal to the 6DT6 can drop to as low as ⅓ volt rms at the secondary winding of the detector input transformer without a loss of oscillation. Normally, the signal generated in the control-grid circuit by the oscillation is approximately one volt rms. The space-charge coupling to the quadrature grid is accompanied by a voltage gain. The oscillations in the quadrature tank have about three times the amplitude of the oscillations in the input circuit. This amplitude is sufficient to develop the required bias voltage across the 560K resistor in the quadrature circuit.

The locked feature of the locked-oscillator circuit refers to the fact that the phase of the oscillations in the control-grid circuit will follow the phase of the incoming sound if signal. During the locked-oscillator mode of operation, the input signal serves as not much more than a type of sync signal having little amplitude but yielding frequency information. Normal quadrature-grid detection takes place in the oscillating detector, and the process is kept under control by the input signal at all times. The oscillation boosts the weak-signal sensitivity of the quadrature-grid circuit, so that its performance becomes comparable to that of a ratio detector. Clear sound can be received even when the station signal is so weak that the picture is not fit to watch.

Limiting of strong signals is done somewhat differently in the locked-oscillator detector than in the gated-beam detector. The characteristic curve of the control-grid voltage of the 6DT6 does not show the rapid leveling off or saturation of plate current at small positive values of grid voltage as the corresponding curve for the 6BN6 does. Limiting in the 6DT6 depends on the damping of strong signals in the grid circuit. A strong input signal causes the control grid to draw considerable current, which loads down the tuned circuit connected to the grid. The oscillation is suppressed by the grid loading, the tuning of the input circuit is broadened, and the peak voltage swing at the grid is held to only a few volts.

Degeneration of audio-frequency signals in the cathode circuit of the locked-oscillator detector also contributes to limiting. The value of cathode resistance required in this circuit for best am re-

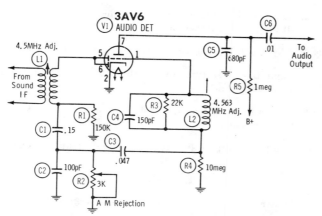

Fig. 14-11. The delta sound detector.

jection is not critical, and no control is needed in the cathode circuit.

The Delta Sound Detector

Fig. 14-11 shows the delta sound detector circuit, which uses a dual-diode/triode tube. The fm sound signal is demodulated in this circuit by a simple, well-known process called slope detection. The process has seldom been applied to commercial circuits because it cannot reject am noise interference. The delta design makes the slope detector practical by coupling the detector with an efficient noise-suppression circuit.

A frequency-modulated, 4.5-MHz signal fed into the secondary of L1 from the sound if stage is developed across C2, which is part of the tuned circuit associated with the secondary of L1. From C2, the signal is coupled through C3 to the tuned circuit composed of L2, C4, and R3. This tuned circuit, pulsed by the signal fed to it through C3, develops the actual driving signal for the grid of the triode. If the pulsing signal is negative with respect to ground at a certain instant, the voltage applied to the triode grid (connected to the other end of the tuned circuit) is positive at the same instant.

The circuit of L2, C4, and R3 is not tuned to exactly 4.5 MHz, as in many sound detectors, but to 4.563 MHz. The result is shown in Fig. 14-12. The tuned-circuit response in other sound detectors falls off symmetrically on either side of 4.5 MHz, whereas the entire bandpass of the sound if signal is on one slope of the response curve of the delta circuit—hence, the name slope detector. The amplitude of the signal pulses developed by the tuned circuit and impressed on the triode grid will change linearly as the incoming signal frequency varies. This change is shown in Fig. 14-12. Thus, any frequency modulation in the input signal will develop a corresponding amplitude modulation in the signal at the grid of the triode. The action of the triode is such that it rectifies and amplifies this signal. In the plate circuit, capacitor C5 filters the 4.5-MHz component of the signal to ground, allowing the audio signal to pass on to the audio output stage.

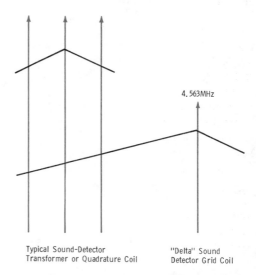

Fig. 14-12. Frequency response of the delta sound detector.

No amplitude modulation should be present in the signal output from the triode except the amplitude modulation due to slope detection. The diode circuit and the am rejection control act to reduce any am interference signals. The average conduction of the diodes changes whenever an amplitude change occurs in the input signal from the preceding stage. The change in average current through the secondary of transformer L1 is reflected back to the plate of the preceding stage as a changing load impedance. This causes a compressing action on any am peaks which might appear in the signal. The residual am peaks appear across R2 as negative peaks only. When these negative peaks are applied to the grid of the triode through C3 and the tuned circuit, they are prevented from reaching the plate circuit, because the tube is biased near cutoff by the proper setting of control R2.

QUESTIONS

1. What is the standard sound if frequency for television receivers?

2. What is the deviation of the television sound carrier for maximum modulation?

3. In a black-and-white television receiver, the sound signal may be taken off at the outputs of what two stages?

4. Where must the sound signal be taken off in a color television receiver?

5. What sound-detector circuit is commonly used in transistor television receivers?

6. In the ratio-detector, are the diodes arranged in opposition to one another or in series?

7. Why is a limiter unnecessary with a ratio detector?

8. Name the three functions of the 6BN6 used in a gated-beam detector.

9. How is limiting accomplished in the locked-oscillator detector circuit?

10. What is meant by an integrated-circuit sound system?

EXERCISES

1. Draw a basic sound system using an integrated circuit.

2. Draw a basic ratio-detector circuit.

CHAPTER 15

Video Amplifiers

Although the picture tube normally requires a grid swing of approximately 40 volts for its range from black to white, the output of the video detector seldom exceeds a few volts. Therefore, the signal from the video detector must be processed through one or more stages of amplification.

In our study of the nature of the video modulating signal, we have seen that the range of frequencies extends from 30 Hz to over 4 MHz. For a video amplifier to provide uniform gain over this extended range, compensating circuits must be used. The basic circuit to which these correction networks are applied is the familiar RC-coupled audio amplifier.

BANDWIDTH AND GAIN
CHARACTERISTICS

A typical RC-coupled audio amplifier with its coupling elements is shown in Fig. 15-1A. The interelectrode capacitances of the tube are indicated by the dotted lines. When this amplifier is employed as a conventional audio amplifier, the solid-line curve of Fig. 15-1B would represent an adequate gain-versus-frequency characteristic for sound reproduction. The dotted line in Fig. 15-1B shows the required gain characteristics of an amplifier for the video band of frequencies.

In order to extend the range of the RC-coupled amplifier shown in Fig. 15-1A, various methods of low-frequency and high-frequency compensation must be employed. Low-frequency compensation is used to overcome the effects of the RC-coupling components, while high-frequency compensation is used to overcome the effects of the total circuit capacitance (shunt capacitance). Regions 1 and 3 in Fig. 15-1B represent the extended frequency range afforded by these compensation networks.

Fig. 15-2 shows the effect of changing the value of the plate-load resistor in an RC-coupled amplifier stage employing a high-transconductance pentode. The band of video frequencies over which the output is flat is greatly extended as the plate-load resistance is decreased. The narrower frequency response with higher values of load resistance is due to the stray circuit capacitance which shunts the load resistor. This shunt capacitance causes the load impedance to be reduced at higher frequencies, resulting in a smaller voltage drop across the load resistor. Therefore, the choice of load resistance is a compromise between bandwidth and gain. The voltage gain of a video amplifier is seldom more than 20, while gains as high as 150 are common in RC-coupled audio-amplifier stages. After the value of load resistance is determined, the stage is compensated to raise the gain at frequencies below approximately 100 Hz and above several hundred kilohertz.

A two-stage transistor video amplifier is shown in Fig. 15-3. The first stage is an emitter-follower configuration in order to match the high-impedance output of the video detector. The 1200-ohm emitter-load resistor provides sufficient wideband response in this circuit. The common-emitter video output stage employs a 6800-ohm collector-load resistor and, therefore, requires high-frequency compensation.

LOW-FREQUENCY COMPENSATION

The low-frequency components of the video signal are generated by scanning large objects or large areas of uniform tone. Poor low-frequency response produces improper contrast of large areas to the smaller objects or fine detail of the picture.

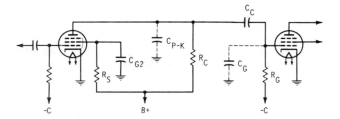

(A) Typical RC-coupled audio amplifier stage.

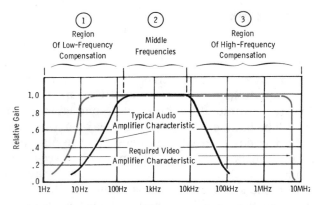

(B) Required video bandwidth compared with typical audio bandwidth.

Fig. 15-1. Response requirements of audio amplifiers compared to that of video amplifiers.

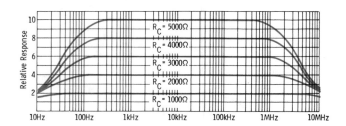

Fig. 15-2. The effect of the plate-load resistor on the gain and bandwidth.

The frequency response of the circuit shown in Fig. 15-1 drops off at the low-frequency end of the video range (region 1 in Fig. 15-1B). Coupling capacitor C_C in series with grid resistor R_G acts as a voltage divider. As the frequency decreases, the reactance of capacitor C_C increases. Since the

voltage across grid resistor R_G constitutes the output voltage of the amplifier, the output drops as the capacitive reactance increases.

To compensate for this voltage drop, a "bass-boost" network can be used in the plate-load circuit as shown in Fig. 15-4A. This bass-boost network consists of plate-load resistors R3 and R4 and capacitor C4. The bass-boost circuit increases the plate load as the frequency decreases. The size of capacitor C4, usually an electrolytic, is such that at all frequencies above approximately 100 Hz its reactance is low compared with that of resistor R4. Capacitor C4 virtually short circuits the resistor, leaving load resistor R3 effective for the middle- and high-frequency range. As the frequency decreases, the reactance of C4 increases. This increases the total plate load and the stage gain increases, as shown in Fig. 15-2.

Fig. 15-4C shows the effect on the low-frequency gain when the time constant of R4 and C4 is changed. In the circuit in Fig. 15-4A, the time constant of R4 and C4 is one second.

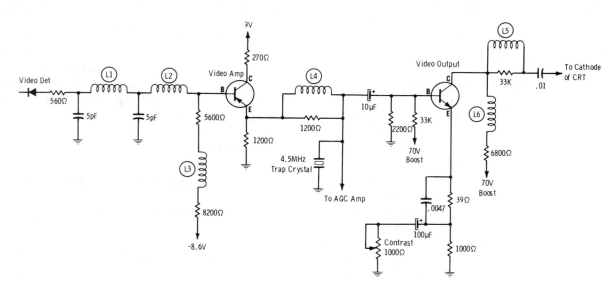

Fig. 15-3. A transistor video-amplifier circuit.

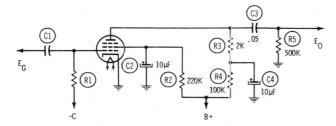

(A) Low-frequency compensation for effect of coupling network C3-R5.

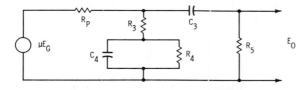

(B) Equivalent low-frequency circuit.

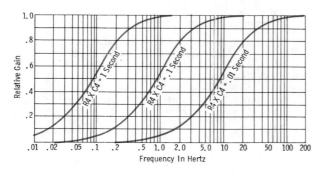

(C) Effect of correction network R4-C4 on gain.

Fig. 15-4. Low-frequency compensation for loss of gain due to coupling circuits.

Other sources of attenuation or loss of low-frequency gain in the RC-coupled amplifier of Fig. 15-1A are the screen and cathode circuits. The effect of the screen circuit can be minimized by the use of a large electrolytic bypass capacitor. When bias for the video-amplifier stage is obtained from a resistor in the cathode circuit, the resistor can introduce degeneration and reduce the stage gain unless it is bypassed by a sufficiently large capacitor. The reactance of this capacitor increases as the frequency decreases, causing a loss of low-frequency gain. Low-frequency loss from this source can be compensated in the same manner as described for the coupling network. To avoid this loss, many receivers operate with the cathode grounded and bias voltage is developed by grid current through a high value of grid-leak resistor.

Poor low-frequency response is much less of a problem in later-model television receivers. Since most tube-type, black-and-white sets use only a single stage of video amplification, there is no low-frequency loss due to the RC-coupling components. Although most tube-type color receivers use two stages of video amplification, the first stage is usually a cathode-follower configuration. The output signal is developed across the relatively low impedance of the cathode circuit, which results in adequate low-frequency response.

Most transistor television receivers employ two or more stages of video amplification. Often, direct coupling is used between transistor video-amplifier stages. However, as shown in Fig. 15-3, RC coupling is used in many transistor video amplifiers. Due to the low impedances of transistor circuits, comparatively large-value capacitors can be used for coupling. The longer time constant of the RC-coupling networks employing these larger capacitors minimizes the loss of low-frequency response.

HIGH-FREQUENCY COMPENSATION

The high-frequency region of video amplification (region 3 in Fig. 15-1B) is responsible for the fine detail in the reproduced picture. If the gain of the video amplifier is deficient at high frequencies, fine vertical lines or small picture elements will be blurred or even missing. Lack of high-frequency response can be detected by examining the narrow portion of the vertical wedges in the test pattern. If a standard test pattern is not readily available, a good crosshatch pattern can be used to judge the high-frequency response. Sharp, well-defined, vertical lines indicate that the high-frequency response is adequate.

Loss of high-frequency gain in an RC-coupled amplifier is caused by the shunting effect of the stray capacitance of the various circuit elements to ground. In a vacuum-tube circuit, these stray capacitances include the plate-to-cathode capacitance of the amplifying tube, the capacitance of the various coupling elements (resistors and capacitors), and the input capacitance of the next tube. In transistor stages, the stray capacitances include the junction capacitance of the transistors and the capacitance of the associated circuit elements.

We have seen in Fig. 15-2 that the high-frequency response of a tube-type amplifier can be extended by decreasing the value of the plate-load resistor. This is accomplished at the expense of stage gain. The gain of the stage at high frequencies is proportional to the parallel combination of the plate-load resistor, the grid resistor of the next stage, and the total shunt capacitance of the circuit. When the capacitive reactance of the shunt capacitance is equal to the value of the resistors in parallel, the gain of the stage is 70.7% of its flat, middle-frequency value. The frequency at which this occurs is shown at point A on the uncompensated curve in Fig. 15-5C. This frequency is usually considered the limit at which the gain can be brought up to mid-frequency level by corrective networks. These networks employ inductors (known as peaking coils) whose reactances increase with increasing frequency and thus compensate for the loss of impedance due to the shunt capacitance. The basic compensation networks employed in television receivers are shunt peaking and series peaking. Most receivers use a combination of series and shunt compensation.

Shunt Peaking

The basic shunt-peaking type of high-frequency compensation is shown in Fig. 15-5A. Although peaking coil L is in series with plate-load resistor R_C, it raises the impedance of the shunt circuit consisting of the stray capacitance (C_S) and the plate-load resistor. Fig. 15-5B shows the equivalent circuit which is in parallel, or shunt, with the plate resistance and the grid resistor of the following stage. The parallel-resonant circuit consisting of L, R_C, and C_S approaches resonance at the high-frequency end of the video band, causing the impedance of the network to increase. Due to the increased shunt impedance at these high frequencies, the high-frequency response of the circuit is improved.

Curve 1 of Fig. 15-5C shows the frequency characteristic of the circuit in Fig. 15-5A before correction is applied. Notice that the 70% response point (point A) occurs at approximately 3.5 MHz. At this point, the capacitive reactance of C_S equals the effective shunt resistance of R_C and R_G in parallel. At this frequency, the value of L is chosen so that the parallel impedance of the circuit consisting of R_C, R_G, C_S, and L is increased to yield the same gain as obtained in the middle-frequency range. This condition is shown as curve 2 in Fig. 15-5C.

The network in Fig. 15-5B can be recognized as a parallel-tuned circuit with series and shunt resistors. Under the condition for correct compen-

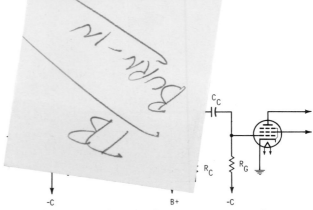

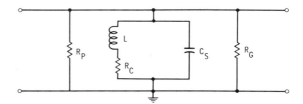

(A) Shunt peaking, high-frequency compensation.

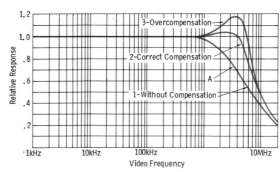

(B) Equivalent high-frequency circuit.

(C) Effect of shunt-peaking choke.

Fig. 15-5. High-frequency compensation by shunt peaking.

sation (curve 2 in Fig. 15-5C), the resonant frequency of the circuit is 1.41 times the frequency discussed in the previous paragraph. If the inductance is increased to reduce the resonant frequency, overcompensation will be obtained, as shown in curve 3 in Fig. 15-5C. Such overcompensation will make fine details in the picture too dark with respect to large areas.

A transistor video-amplifier stage utilizing shunt peaking is shown in Fig. 15-6. The collector

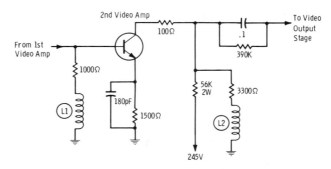

Fig. 15-6. Shunt peaking in a transistor video amplifier circuit.

load for this circuit is the 3300-ohm resistor in series with peaking coil L2. The 56K collector-supply resistor is shunted by the relatively low impedance of the collector load. Peaking coil L2 and the 3300-ohm resistor are effectively in parallel with the stray capacitance of the circuit. Therefore, the effect of peaking coil L2 is the same as the peaking coil in the vacuum-tube circuit in Fig. 15-5A. Additional high-frequency compensation is provided by peaking coil L1 in the base circuit. The inductance of peaking coil L1 is much smaller than that of peaking coil L2 due to the comparatively large capacitance of the forward-biased emitter-base junction and the fact that a circuit with a larger capacitance requires a smaller inductance to be resonant at a given frequency. The peaking coils used in television receivers are designed to operate with particular types of tubes or transistors and associated circuit capacitances. Therefore, when it is necessary to replace peaking coils, exact replacements must be used.

Series Peaking

Fig. 15-7A shows another method of raising the response curve at the high-frequency end. The peaking coil in series with coupling capacitor C_C forms a series-resonant circuit. The resonant frequency of this circuit is made somewhat higher than the upper limit of the desired video range. The decreased impedance of the coupling circuit at higher frequencies counteracts the loss due to the lower reactance of the plate-to-cathode capacitance across the load resistor. This permits the use of a higher value of load resistance with consequently higher stage gain. The gain of a video stage with series peaking can be made 50% greater than one with shunt peaking alone for the same bandwidth. Series peaking is used in the same manner for transistor video-amplifier stages, as shown in Fig. 15-3.

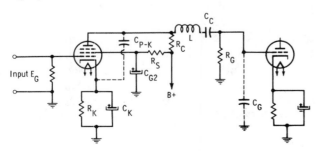

Fig. 15-7. High-frequency compensation by series peaking.

Combination Series and Shunt Peaking

From the foregoing discussion, it is evident that shunt peaking and series peaking operate independently and may be used to complement each other. Many modern television receivers employ a com-

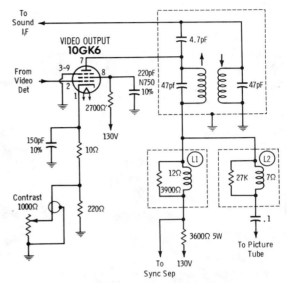

(A) Typical vacuum-tube configuration.

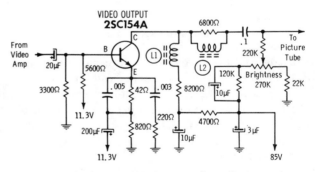

(B) Typical transistor configuration.

Fig. 15-8. Video stages using combination series and shunt peaking.

bination of both methods. Fig. 15-8 shows examples of vacuum-tube and transistor video-amplifier stages using this combination method for high frequency peaking. In both of these circuits, inductor L1 operates as a shunt-peaking inductor, and inductor L2 operates as a series-peaking inductor.

Note that all of these coils (with the exception of L1 in the transistor circuit) have a shunting resistor connected across them. The shunting resistor reduces the Q of the circuit to flatten out the gain-versus-frequency curve. This resistor also damps the circuit and prevents it from being shock excited into transient oscillation by sharp video pulses or noise peaks. Such oscillation produces smearing and a negative image following the fine detail in the picture, as shown in Fig. 15-9. No damping resistor is required across L1 in Fig. 15-8B because sufficient shunt resistance is already present in the circuit.

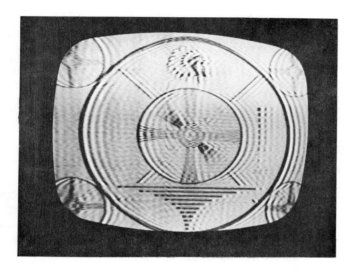

Fig. 15-9. A test pattern showing the effect of transient oscillation.

In the discussion of video detectors in Chapter 13, mention was made of series and shunt peaking for correction of high-frequency losses in the detector-output circuit. The peaking method used in these circuits is the same as the type just discussed.

PHASE SHIFT IN THE VIDEO AMPLIFIER

In our discussion of the video amplifier up to this point, we have only considered the requirement of gain versus frequency. Another equally important consideration is the possible phase shift, or time delay, of the signal as it passes through the amplifier.

In the amplification of sound waves in a broadcast receiver, phonograph amplifier, or sound system, phase shift at one end of the audio-frequency spectrum with respect to the other is seldom important. The ear is not sensitive to phase shift, and, for this reason, the service technician need not give it any consideration. In the television video amplifier, however, phase shift is extremely

important. If not corrected, phase shift can cause badly distorted and smeared pictures.

Almost all sound waves are sinusoidal. An understanding of the action of voltage sine waves passing through an amplifier provides a sufficient background for understanding the operation of sound amplifiers. The video signal, however, is often a square or rectangular waveform. This can readily be understood by considering the output from the camera tube as it scans a black bar in front of a white background. During the scanning process, the video signal is near its zero amplitude as it crosses the background. Suddenly, the video signal rises to maximum amplitude and remains there until the bar is crossed; then it drops to zero amplitude again when the background is reached. The width of the bar determines the half wave of a low frequency. This frequency is the fundamental frequency for that particular element of the picture. The square wave video signal is composed of this fundamental frequency plus a great number of harmonics of different amplitudes. Together, this fundamental frequency and the harmonics will produce the rapid rise and fall at the ends of the bar. Thus, to produce a black bar, no matter what its length, all of the harmonic frequencies must be amplified. As we have previously seen, these harmonic frequencies can extend up to 4 MHz.

In our study of the action of amplifier stages, we learned that the wave applied to the grid of a vacuum stage is shifted 180° in phase when it appears across the plate-load resistor. In a like manner, a wave applied to the base of a transistor stage is shifted 180° as it appears across the collector load. As far as the tube or transistor itself is concerned, a wave of any frequency in the video range is shifted 180° in phase as it passes through a single stage. The network of resistors and capacitors which constitutes the coupling elements between the stages can, however, cause a phase shift which differs both in amount (number of degrees) and in direction for different frequencies in the video range. Note that a cathode-follower or emitter-follower configuration is sometimes employed as one of the stages in a video amplifier. In these stages, of course, there is no phase inversion in the tube or transistor itself.

For the middle range of frequencies (region 2 in Fig. 15-1B), the coupling network is resistive, and a constant 180° phase shift occurs because of the action of the tube or transistor.

Low-Frequency Phase Shift

At low frequencies (region 1 in Fig. 15-1B), the RC coupling network produces a leading phase shift which increases as the frequency decreases. This phase shift is proportional to the ratio of the reactance of the coupling capacitor to the resistance of the grid resistor. If no correction were applied, this phase shift would cause the effects shown in the table in Fig. 15-10. A very small phase shift can cause a large time difference at very low frequencies. As a result, large areas will show smearing at the edges and an uneven tonal reproduction in the picture. Excessive phase shift at very low frequencies (30 to 70 Hz) will also cause a gradual shading of the picture from top to bottom, because the effect of a single picture element lasts for more than one horizontal line.

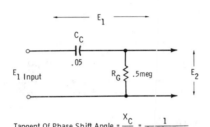

Tangent Of Phase Shift Angle = $\dfrac{X_C}{R_G} = \dfrac{1}{2\pi F C_C R_G}$

Frequency (Hz)	Phase Shift of E_2 (Degrees)	Time Delay (μs)	Horizontal Displacement of Image (Inches) (21-in. Tube)
500	0.7	4	1.4
200	1.8	25	9.0
100	3.6	100	1 Line +13
70	5.2	206	3 Lines + 5.5
50	7.2	400	6 Lines + 6.8
30	12.0	1110	17 Lines + 10.8

Fig. 15-10. Table showing low-frequency phase shift due to coupling elements.

Since the phase shift is a leading effect, it can be corrected by a shunt circuit consisting of a capacitance and resistance in parallel. This is the same type of network required to compensate for the loss of low-frequency gain (R4-C4 in Fig. 15-4). Thus, both phase shift and loss of low-frequency response can be corrected by the same network.

High-Frequency Phase Shift

At high frequencies (region 3 in Fig. 15-1B), shunt capacitance C_S can cause a lagging phase shift of the high frequencies with respect to the middle-frequency portion of the video signal. Fig. 15-11 shows those parts of the equivalent amplifier circuit responsible for high-frequency phase shift. The table shows that while the phase shift increases, the time delay drops, and the image displacement becomes less. Again, as with low-

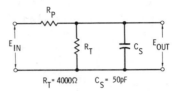

$R_T = 4000\Omega$ $C_S = 50\text{pF}$

Tangent Of Phase Shift Angle = $-2\pi F C_S R_T$

Frequency	Phase Shift (Degrees)	Time Delay (μs)	Horizontal Displacement Of Image (Inches) (21-in. Tube)
10kHz	.72	2.0	.71
100kHz	7.2	2.0	.71
1MHz	64.4	1.8	.64
2MHz	103.8	1.4	.5
3MHz	124.0	1.266	.45
4MHz	136.6	0.95	.34

Fig. 15-11. Table showing high-frequency phase shift due to shunt capacitance.

frequency phase shift, the same corrective networks which compensate for loss in gain are used to produce a corrective phase shift. The shunt- and series-peaking coils and their combinations produce an overall circuit phase shift which is proportional to frequency.

Ideal Phase Shift for Uniform Time Delay

At the horizontal scanning frequency (15,750 Hz), the spot moves across a picture tube 19-inches wide at approximately 356,000 inches per second. Thus, a time delay of one microsecond will produce a difference in position of about three-eighths of an inch. If all frequencies in the video signal were subject to this same time delay, the picture would be satisfactory. Although the picture would be displaced three-eighths of an inch to the right, it could be moved back by adjusting the horizontal centering.

In order to have a uniform time delay, the phase shift must be different for each frequency (a phase shift proportional to frequency. The relationship between time delay and frequency is expressed by the equation:

$$\text{Time Delay} = \frac{\text{Phase Shift in Degrees}}{360° \times \text{Frequency in Hertz}}$$

Fig. 15-12 shows the desired phase-shift relationship proportional to frequency, or uniform time delay, as applied to the video amplifier of a television receiver. At high frequencies, the phase shift of the uncompensated amplifier drops below the desired proportional curve which will produce a picture without blurred or shifted elements.

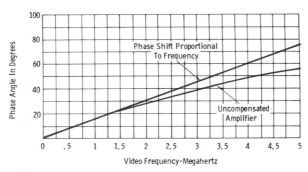

Fig. 15-12. Phase-shift requirements of a video amplifier for perfect reproduction.

SMEARING OF THE PICTURE DUE TO AMPLITUDE OR PHASE DISTORTION

Fig. 15-13 illustrates the effects on a square wave when various deficiencies exist in the video amplifier. Fig. 15-13A shows the correct video signal that should be produced when the black bar is scanned.

Fig. 15-13B shows the square waveform after it has passed through an amplifier having insufficient high-frequency gain. The leading edge of the amplified wave has a gradual slope, rather than an abrupt rise. This slope will cause shading from gray to black in the reproduction. At the trailing edge, the exponential decay of the wave will cause a gray-to-white smear.

Fig. 15-13C shows the effect of overcompensation, or excessive low-frequency response. This effect is the same as the effect caused by insufficient high-frequency response, although not as pronounced.

Fig. 15-11D illustrates insufficient low-frequency response accompanied by attendant phase shift. The front edge of the bar is black, shading to gray. The smear following the bar is white, shading to gray. This type of smear is sometimes called a "trailer."

The effect of insufficient damping in the peaking circuits used for high-frequency compensation

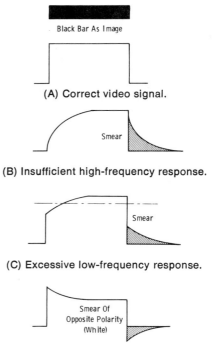

(A) Correct video signal.

(B) Insufficient high-frequency response.

(C) Excessive low-frequency response.

(D) Insufficient low-frequency response.

Fig. 15-13. Square-wave video signals showing the effects of amplifier deficiencies.

was shown in Fig. 15-9. The oscillations due to shock excitation of these peaking circuits by the short-duration square-wave picture elements can be seen as alternate white and gray ghosts adjacent to the black circular rings of the test pattern.

BRIGHTNESS CONTROL

Our previous study of the television receiver system concerned itself with the motion of the electron beam in the picture tube as it scanned the picture area. We showed in Chapter 1 that the intensity of illumination of the picture-tube screen could be controlled by a grid element in the gun of the picture tube. At several points in the text, we have indicated that the variation of the video signal on this control grid is responsible for the instantaneous changes of spot illumination which make up the elements of the picture. When the signal voltage on this control element changes in the negative direction, a darker spot is produced on the screen. Finally, at some critical negative voltage, the spot of light is entirely extinguished. We have also shown that a video signal with opposite polarity can be applied to the cathode with the same results.

One of the essential controls for a television receiver is a bias adjustment for the picture tube. The adjustment ensures that the blanking level, or pedestal, of the signal occurs at the black point. This bias adjustment is the brightness control. Fig. 15-14 shows two brightness-control circuits which establish the bias voltage for the control grid with respect to the cathode and thus determine the correct picture brightness. In Fig. 15-14A, the video signal is applied to the cathode, and the setting of the brightness control determines the cathode voltage. This is the configuration found in most modern black-and-white tube-type receivers.

The circuit in Fig. 15-14B shows the video signal applied to the control grid of the picture tube. In either circuit, the brightness control adjusts the bias voltage between the grid and cathode and establishes the correct blanking or black level. When direct coupling is employed between the video amplifier and the picture tube, the brightness control may be located in any of the direct-coupled video-amplifier stages. In most color receivers, the video signal is direct coupled to the cathode of the picture tube, and the brightness

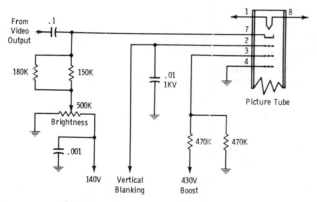

(A) Video signal applied to cathode.

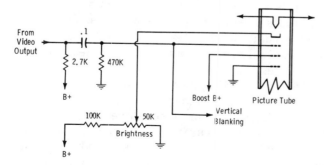

(B) Video signal applied to control grid.

Fig. 15-14. Brightness-control circuits.

control is usually located in the grid circuit of the video-output stage.

CONTRAST CONTROL

The contrast control of a television receiver has about the same function as the volume control of a radio receiver or a stereo amplifier. The "contrast" of the picture is the ratio of light intensity of the brightest highlights in the picture to that of the darkest portion of the picture. Because of scattering of light at the fluorescent face of the picture tube, the black portion of the picture can never be an absolute black, or total absence of light. Since the ratio of maximum illumination to the illumination at the cutoff point is directly proportional to the voltage swing of the video signal applied to the picture tube, the contrast can be controlled by varying the output of the video amplifier. The output can also be controlled in the video if amplifier, as was done in early tv receivers.

Fig. 15-15 shows two methods of contrast control by varying the load resistance of a stage. In some receivers, the contrast control is a potentiometer employed as a load resistor for the output of the video driver as shown in Fig. 15-15A. In this circuit, the setting of the contrast control determines the amount of video signal applied to the video-output stage.

Another arrangement used in solid-state receivers is shown in Fig. 15-15B. Here, the contrast control is a variable load resistance in the collector circuit of the video-output transistor, designed to directly control the amount of video signal fed to the picture tube.

Fig. 15-16 shows one of the most popular methods of contrast control whereby a variable resistor in the video-output circuit is used to change the gain of the stage. This arrangement is used to a great extent in both tube-type and transistor television receivers.

In the vacuum-tube circuit shown in Fig. 15-16A, the contrast control is connected between the cathode of the video-output tube and ground. In this particular circuit, the grid resistor is returned to the cathode end of the control. Therefore, the bias on the tube is not varied when the control is reset. Instead, the effective B+ voltage on the tube is varied when the control is adjusted. The grid resistor is connected to ground in many circuits of this type. To avoid changing the dc bias

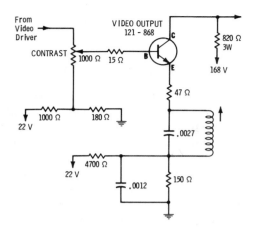

(A) Varying the input to the video-output stage.

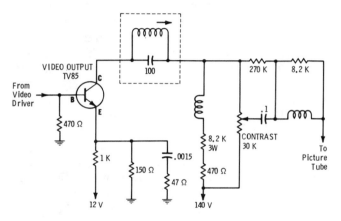

(B) Varying the output from the video-output stage.

Fig. 15-15. Contrast-control circuits.

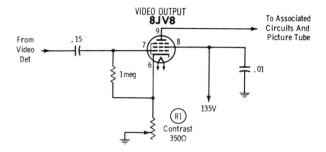

(A) Typical vacuum-tube configuration.

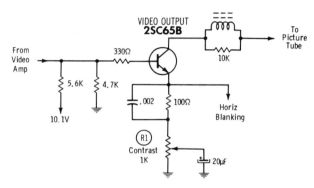

(B) Typical transistor configuration.

Fig. 15-16. Method of contrast control by varying the gain of the video-output stage.

on the tube, the center arm of the contrast control is connected to ground through a large-value electrolytic capacitor. The lower end of the contrast control is also connected to ground. Therefore, the amount of cathode-circuit resistance which is bypassed is determined by the setting of the contrast control. The transistor counterpart of this arrangement is shown in Fig. 15-16B.

VIDEO COUPLING TO CRT

The signals from the video-output stages shown in Fig. 15-17 are ac coupled to the cathode of the picture tube through a .1-µF capacitor. In these circuits, the level of the dc voltage on the cathode is determined by the setting of the brightness control. When the video signal is ac-coupled to the cathode, the average value of the video signal always coincides with the preset dc level as shown in Fig. 15-18. Theoretically, this is not a workable

arrangement because the composite video signal is supposed to play some part in the establishment of the dc voltage on the cathode. Regardless of the shifts which may occur in the average value of the video signal, the pedestals of the sync pulses in that signal should remain at a constant dc level equivalent to the cutoff voltage of the picture tube. In this manner, the pedestals of the sync pulses establish a reference point for all black portions of the picture.

When the video signal is allowed to arrange itself around a fixed average value, as is the case in an ac-coupled amplifier, the level of the sync pedestals shifts somewhat whenever the average brightness level of the picture changes. Therefore, a given shade of gray is not reproduced exactly the same in a predominantly dark scene as it is in a light scene.

In actual practice, a satisfactory picture can be obtained with an ac-coupled video amplifier and many modern television receivers employ this type of circuit. The appearance of retrace lines in the picture, which is a serious disadvantage of ac coupling, is prevented by the use of an efficient retrace-blanking circuit. Sharp negative pulses with an amplitude of more than 50 volts are ob-

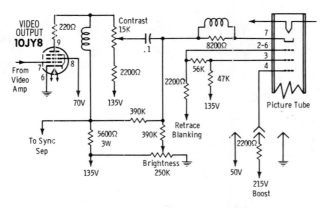

(A) Typical vacuum-tube configuration.

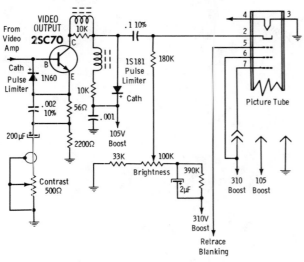

(B) Typical transistor configuration.

Fig. 15-17. Single-stage video amplifiers employing ac coupling to the picture tube.

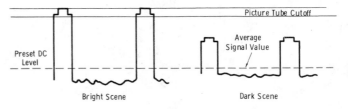

Fig. 15-18. Average value of video signal coincides with preset dc level.

tained from the vertical-deflection circuits and applied to the grid of the picture tube. These blanking pulses cut off the picture tube during the retrace period so that the retrace lines do not appear on the screen.

Many early television receivers employed a diode which functioned as a dc restorer. This diode re-established the fixed black level that was lost in capacitive coupling. The dc-restorer diode was generally used in two-stage video amplifiers where the video signal was ac coupled to the control grid of the picture tube. A portion of the resistive path between the grid and ground was shunted by the diode. The conduction of the diode depended on the peak value of the sync-pulse tips in the video signal, and a capacitor was charged in proportion to the amount of conduction by the diode. The charge on this capacitor regulated the dc-voltage level on the grid of the picture tube and caused the pedestals of the sync pulses to always be held at a constant level. The dc restorer ensured that all signals representing black objects would drive the picture tube to cutoff if the brightness control was correctly set.

QUESTIONS

1. What happens to the picture when the video amplifier has poor low-frequency response?

2. How is the effect of the screen circuit on low-frequency response minimized in tube-type receivers?

3. What happens to the picture when the video amplifier has poor high-frequency response?

4. What causes the loss of high-frequency response in the video amplifier? How is this loss minimized?

5. What is the function of the brightness control?

6. Name three ways the contrast control can be used in the video-amplifier circuit to control the amplitude of the video signal applied to the picture tube.

7. When the video signal is ac coupled to the cathode of the picture tube, and the grid is effectively grounded, what determines the level of dc voltage on the cathode?

EXERCISES

1. Draw the basic circuit of a video amplifier with shunt peaking.

2. Draw the basic circuit of a video amplifier with series peaking.

CHAPTER 16

Automatic Gain Control

Automatic gain control (agc) minimizes the effect of changes in signal strength at the receiver antenna. The gains of the rf and if stages are so regulated that a strong signal is amplified less than a weak signal. As a result, the quality of the tv picture tends to be relatively constant.

Variations in signal strength are of two types: (1) variations between signals received on different channels, and (2) variations occurring from time to time on the same channel.

Both strong and weak channels are available in many locations. When agc is provided in the receiver, the contrast control does not need to be reset each time a new channel is tuned in. Agc also compensates for extremely strong signals received from powerful stations.

The agc system levels out most of the periodic amplitude variations which would cause fading on a particular channel; therefore, a steady picture is obtained, even in moderate fringe areas. The rapid flutter caused by airplanes flying near the path of the transmitted signal is also corrected. as much as possible, through agc action.

RECTIFIED AGC

An agc circuit composed of a simple filter network can be seen in the schematic in Fig. 16-1A. The filter is composed of R4 and C1. The time constant of this filter circuit is approximately 0.15 second.

The performance of this system can be refined by the use of a special agc diode and a charging time constant shorter than the discharging time constant. A circuit with these modifications will not be affected by variations in scene brightness.

Such variations cause shifts in the amplitude of the carrier in the television signal.

The average value of voltage of a video signal is obviously not an absolutely true indication of signal strength. If the agc voltage were developed from this average voltage, the agc filter output would tend to increase during the transmission of scenes containing many large, dark objects or a dim background. Although the amplitude of the video signal for a bright scene is greater than that for a dark scene, the average voltage value for the dark scene is greater. Since the dc component is still present at the video-detector output, the blanking pedestals will occur at a fixed point in reference to the zero dc level. The average voltage for the dark scene will be nearer the blanking level and, therefore, will occur at a greater distance from the dc level as shown in Fig. 16-2.

The only portion of the composite video signal which has a constant amplitude regardless of picture content is the sync-pulse signal. If the strength of the received signal does not change, a consistent peak value of voltage is reached by the tips of these pulses. This peak is comparable to the maximum voltage attained during 100% modulation of a carrier by an audio signal. Improved agc action will be obtained if the agc filter capacitor can be charged to this peak voltage and if most of this charge can be maintained between pulses.

If the charging and discharging time constants of the agc filter are of nearly equal lengths, the system can never build up a charge approaching the peak amplitude of the signal voltage. The discharging time constant can be lengthened in order

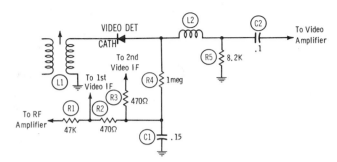

(A) Agc circuit composed of a simple filter network with a time constant of approximately 0.15 second.

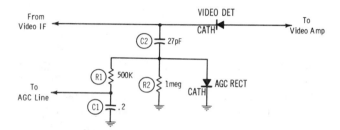

(B) Agc circuit with improved action obtained by a lengthened time constant in the discharge path.

Fig. 16-1. Schematics of simple agc systems.

that a greater charge can be retained on the filter capacitor. This feature has been included in the circuit in Fig. 16-1B.

Resistor R2 in Fig. 16-1B corresponds to resistor R5 in Fig. 16-1A, but the value of the resistor in Fig. 16-1B has been increased to one megohm. As a result, the charging time constant is 0.1 second, but the discharging time constant is increased to 0.3 second.

A separate diode must be used to rectify the agc voltage if a resistor of high value is used in the discharge circuit of the agc filter capacitor. The reason for this requirement will be clear if it is noted in Fig. 16-1A that the voltage applied to the video amplifier is developed across R5. Most of the high-frequency portions of the video signal would be lost if that resistor were large in value

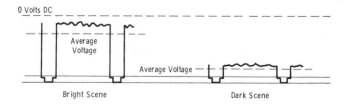

Fig. 16-2. Average voltage values of the video signal for a bright scene and a dark scene.

because the shunting effect of stray capacitance in the video-amplifier input circuit would be exaggerated. Resistor R2 in Fig. 16-1B can be as large as necessary, because the video detector diode is separate from the agc diode.

It should be repeated that agc action can be obtained without special concern for changes of brightness in the picture. However, correcting this condition is important enough that many of the relatively simple agc systems and all of the more complex ones develop the agc voltage from the peak voltage of the sync pulses.

AMPLIFIED AGC

An improvement of the basic agc system was the addition of an amplifier tube to boost the amplitude of the rectified agc voltage. A two-stage, amplified-agc circuit can develop an adequate control voltage when the changes in amplitude of the video signal are so slight that a simple agc system would not respond to them. Amplified agc is, therefore, more efficient than ordinary simple, or filtered, agc.

An amplifier-agc circuit is shown in simplified form in Fig. 16-3. An almost pure dc voltage, positive in polarity, is produced across C1 and R1 when a video signal is applied to the agc rectifier. This signal is directly coupled to the grid of agc amplifier V1.

Since the agc voltage is taken directly from the plate of the agc amplifier, a negative voltage of low amplitude must be present in the plate circuit of V1.

It is assumed that the circuit in Fig. 16-3 can be supplied with a voltage of approximately −100 volts. A voltage equal to about one-half the B− potential is applied to the cathode of the amplifier. The average grid voltage is determined by the set-

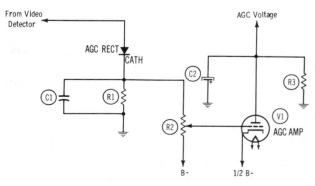

Fig. 16-3. Schematic of an amplified agc circuit.

ting of potentiometer R2. This control is a voltage divider between the full B— voltage and the slightly positive voltage at the cathode terminal of the agc rectifier diode.

The bias on V1 automatically varies in step with the voltage developed across R1 and C1 by the video signal. An increase in amplitude of the incoming sync pulses causes the cathode terminal of the agc rectifier to go in the positive direction; therefore, the voltage at the arm of R2 is less negative than before, and the bias on V1 is reduced. Conduction increases in the amplifier. The plate voltage of V1 swings in a negative direction, and agc filter capacitor C2 in the plate circuit is heavily charged in the polarity shown in Fig. 16-3. The time constant of R3 and C2 is so long that C2 discharges very slowly. The agc control voltage is, therefore, nearly equal to the voltage which charges C2.

The amplified-agc circuit shown in Fig. 16-3 was used in earlier tube-type television receivers. Most late-model tube-type receivers employ keyed agc which will be discussed subsequently. Since the agc keyer tube provides considerable amplification, an additional agc-amplifier stage is not used in most later tube-type receivers. However, agc-amplifier stages are used in many solid-state television receivers that employ keyed agc. The agc-amplifier stages in transistor television receivers are often used to provide impedance matching rather than gain.

The agc voltage used with tube-type circuits is known as *reverse agc*. That is, the agc-control voltage applied to a controlled stage reverse biases it and thereby reduces its gain. Since reverse agc is always used with tube-type circuits, a negative control voltage must be developed by the agc circuits.

The agc-control voltage applied to a transistor stage may be either positive or negative, depending on whether the transistor is an npn or pnp type. Also, many transistor rf and if amplifiers employ *forward agc*. The characteristics of these transistors are such that, when operated with a fairly high resistance in the collector circuit, the gain of the stage is reduced due to saturation when the forward bias is increased. The forward-agc control voltage applied to these stages actually forward biases the transistor in order to reduce its gain.

A transistor agc amplifier is shown in Fig. 16-4. A positive agc voltage from the agc keyer is applied to the base circuit of the transistor through the .0022-μF capacitor. The increased conduction of the transistor causes a positive voltage drop across the 2200-ohm resistor in its emitter circuit. This additional positive voltage is applied to the rf and if amplifiers through their respective agc networks. Since the agc amplifier is an emitter-follower stage, there is actually no voltage gain. However, the stage is an amplifier in the sense that it matches the high output impedance of the agc keyer to the low input impedance of the agc-controlled stages.

KEYED AGC

Keyed agc is the most complex form of automatic gain control used in tv receivers; it is also the most efficient. The circuits in Fig. 16-5 are typical of most circuits of this type. Fig. 16-5A shows a typical agc circuit employed in tube-type

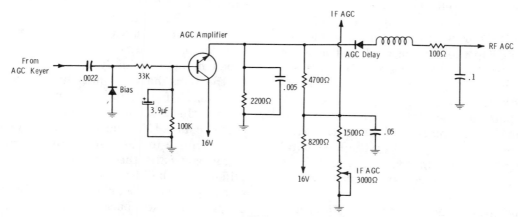

Fig. 16-4. A transistor agc-amplifier circuit.

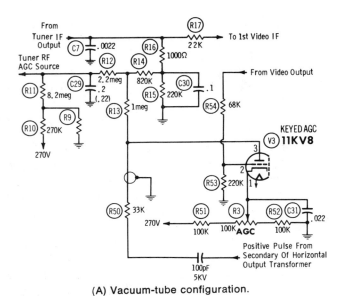

(A) Vacuum-tube configuration.

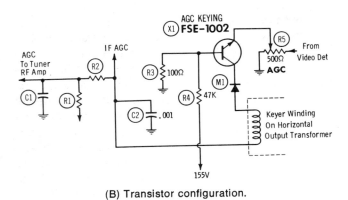

(B) Transistor configuration.

Fig. 16-5. Typical keyed-agc circuit.

receivers, and Fig. 16-5B shows its transistor counterpart. The tube-type circuit will be discussed first.

The agc-keying tube used in this circuit is typically a triode, but it could be a pentode. Two input signals are required. A composite video signal which contains positive-going horizontal-sync pulses is applied to the control grid and positive pulses of high amplitude are coupled to the plate from a winding on the horizontal-output transformer. The conduction of the tube depends on the arrival of a pulse at the plate at the same time that a horizontal-sync pulse arrives at the grid. Since the plate pulses are timed by the horizontal oscillator, they have the same frequency and phase as the sync pulses, if the receiver is correctly synchronized.

A short burst of conduction occurs 15,750 times each second in response to the arrival of each

pulse. Since the level of the sync tips determines the bias on the keying tube during the burst of conduction, the amount of conduction that occurs during the pulses depends on the dc level reached by the tips of the horizontal-sync pulses at the grid. The level of the sync tips, in turn, depends on the strength of the incoming signal. The keying tube is cut off between pulses, and the plate voltage assumes a negative value proportional to the amount of conduction that takes place during the pulses. The greater the conduction, the more negative the average plate voltage. The plate voltage is passed through an RC filter, and the resultant dc voltage is applied to the rf amplifier and to one or more if amplifiers as grid bias.

Since the keying tube is cut off in the interval between horizontal-sync pulses, noise and video in the input signal cannot affect the development of agc voltage. In addition, the keyed-agc circuit does not respond to the vertical-sync pulses. The keying tube conducts only during alternate equalizing and serration pulses. The amount of conduction is not sufficient to cause a periodic rise in agc voltage at the vertical rate of 60 Hz. Such a rise is characteristic of other agc systems. The agc filter in the keyed circuit, unlike the filters in other agc circuits, does not have to be designed to remove vertical pulses from the output voltage. The time constant of the filter is, therefore, appreciably shorter in a keyed-agc circuit than in other agc circuits. This is a desirable feature because it tends to make the keyed circuit largely immune to airplane flutter.

The dc cathode voltage is made several volts more positive than the average dc grid voltage in order that the keying tube will be correctly biased during conduction. The difference between the grid and cathode voltages is more important than their absolute values.

These two voltages have high positive values in most actual circuits. Regardless of this fact, the tube conducts normally because the keying pulses on the plate have a much higher positive value. The reason for the use of a positive grid voltage is that the grid of the keying tube must be directly coupled to the stage supplying an input video signal to the agc system. Frequently, the signal source is the plate circuit of a video amplifier, and the high voltage at the amplifier plate also appears at the grid of the keying tube. Direct coupling is used because the variations in grid voltage, which occur in response to changes in the dc level of the sync pulses, are fundamental to the operation of the agc system. This reference level would be lost if a capacitor were inserted in the input lead to the grid of the keying tube.

As shown in Fig. 16-5A, an isolating resistor (R54) is included in the grid circuit. This resistor limits the amount of current in the circuit if the grid should draw current. The impedance of the resistor also isolates the input capacity of the keying tube from the video circuit; therefore, the loading effect of the agc circuit on the video circuit is greatly reduced.

Recall that the amount of conduction through the tube is determined by the peak amplitude of the sync pulses. If the video signal is weak, the positive voltage on the grid of the keying tube will be relatively low. The difference between the grid voltage and the cathode voltage is great, and the heavily biased tube passes less than average current. Little agc voltage is produced under these conditions because the charge on the capacitors in the plate circuit is comparatively small. On the other hand, a strong video signal develops a high grid voltage which is closer to the value of the cathode voltage. Since the bias on the tube is low, conduction through the tube is heavy and considerable agc voltage is thus produced in the plate circuit.

Fig. 16-5B shows a typical keyed-agc circuit employed in transistor-type receivers. In this circuit, an npn transistor is connected as a common-emitter amplifier. The emitter receives a video signal from the video detector through agc control R5. The base of the transistor is reverse biased at

approximately 3 volts and the collector is pulsed from a winding on the horizontal output transformer. Thus, the transistor only conducts during the flyback interval. Rectification of the pulses through diode M1 produces approximately 5 volts at the collector of X1.

In operation, the video detector applies a negative-going video signal to the emitter. The amount of applied signal can be varied by means of agc control R5. The video signal increases the forward bias on the transistor, causing the collector current to increase. Electrons flow from the collector through diode M1, through the keyer winding, through R2, and through R1. The voltage drop across R1 becomes more negative, and a higher negative bias is therefore applied to the if strip and the rf amplifier in the vhf tuner. C1, R2, and C2 comprise a filter network for the pulse output of keying transistor X1.

Fig. 16-6 shows an agc circuit and a noise inverter combined in one tube with a sync-separator circuit. The sync-separator circuit and noise inverter were discussed in Chapter 10. The input signal for the agc section of the tube is a composite video signal with positive-going sync pulses. The composite video signal is taken from the plate of the video amplifier and applied to one of the second control grids (pin 9). This signal is direct-coupled because the dc level of the sync tips is important to the operation of the agc tube. If the composite video signal is strong, the sync tip level at pin 9 will be less negative than if the video signal were weak. In other words, a strong input signal places a relatively low bias on the left-hand section of the tube. This section of the tube then conducts heavily, and the voltage on the left-hand plate drops. The lower the plate voltage becomes, the more agc voltage is produced.

Since the plate is maintained at a positive potential of approximately 35 volts in order that the tube will conduct, the agc voltage cannot be obtained directly from the plate. Instead, it is derived in the following manner. The plate of the agc tube is connected to the grid of the horizontal-output tube through a voltage divider made up of resistors R3, R2, and R4. The average dc potential at the grid of the horizontal-output tube is −75 volts. Variations in this grid voltage are leveled off by the agc filters. The voltages at the intermediate points on the divider can be applied directly to the agc line. Note that the agc bias voltages for the tuner and for the if strip are taken

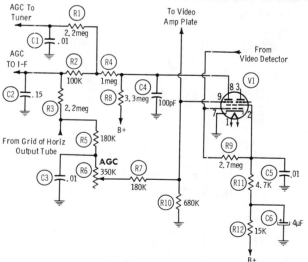

Fig. 16-6. A combination noise-inverter, sync-separator, and agc circuit.

from separate points. When no input signal is being applied to the agc section of the tube, there is a slight positive voltage at the junction of R2 and R4 and a slight negative voltage at the junction of R2 and R3. These voltages are fed to the tuner and if stages, respectively. Both agc voltages are driven in a negative direction when an input signal is applied to the tube, but the if voltage always remains more negative than the tuner voltage. This arrangement amounts to a simple delay circuit for the agc line to the tuner.

The negative potential at the grid of the horizontal-output tube is also utilized by another voltage divider. This voltage divider is in the second control-grid circuit of the agc section of the tube and is composed of R10, R7, agc control R6, and R5. A negative dc voltage is fed from the junction of R10 and R7 to the second control grid of the tube, and the input signal from the video amplifier is superimposed on this dc voltage. The agc control permits the range of bias on the second control grid of the tube to be adjusted.

If noise pulses in the agc signal were uncontrolled, they would cause the agc section of the tube to conduct excessively. The voltages in the agc system would then become too negative. The noise inverter keeps these unwanted pulses from increasing the conduction of the tube through the action of the inverted signal on pin 7. The agc voltages are maintained at normal values even when considerable noise is present in the video signal.

The values of many of the components in the tube circuit are critical. This fact applies especially to the noise inverter. The bias on the inverter grid (pin 7) must be maintained at such a value that the sync tips will almost reach the cutoff voltage of the grid. If the bias is too low, the noise inverter will not accomplish its purpose. If the bias is too high, the tube will be driven into cutoff by each sync pulse. The latter condition would cause low agc voltage and unstable synchronization.

No provision is made for adjusting the noise-inverter circuit. This is actually an advantage because the operation of the noise inverter cannot be upset by misadjustment of some control.

DELAYED AGC FOR RF STAGE

Even the weakest usable input signal develops some agc voltage, which reduces the gain of all stages controlled by the agc system. Unfortunately, the amplitude of a weak signal at the grid of the rf amplifier is barely more than the amplitude of the noise in that stage. When reception is poor, full gain in the rf stage is essential, so that the signal-to-noise ratio can be kept high. Once the signal has been amplified above the level of the noise, it can be acted upon by agc in the if amplifier without undesired effects.

Therefore, provisions are made for delaying the action of the agc bias on the rf amplifier in most receivers employing keyed-agc systems. The delay circuit includes a connection through a high resistance to B+. When a weak signal is being received, the tuner bias is sharply reduced because of this B+ connection. Included in most delay circuits is a clamper diode which conducts and shorts the agc line to ground whenever the bias voltage tends to become positive. The circuit in Fig. 16-7 has this delay feature in the rf branch of the agc line. The components included in the delay circuit are resistors R1 and R6, and clamper diode M1.

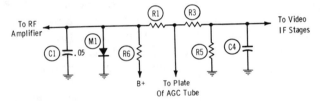

Fig. 16-7. Agc delay circuit.

As shown in the graph of Fig. 16-8, the delay circuit holds the agc bias applied to the rf amplifier to about zero until the incoming signal is strong enough to develop −4 volts of bias in the if section of the agc line. The rf bias appears at this signal level, increases more rapidly, and even-

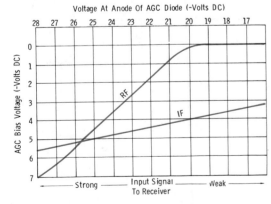

Fig. 16-8. Graph of the rf and if bias voltages when an agc delay circuit is employed.

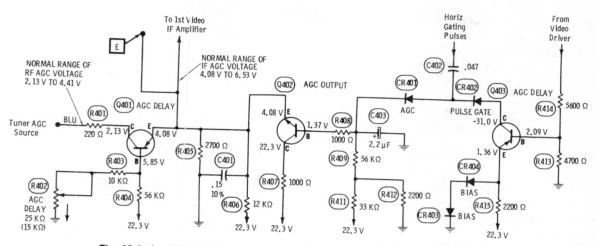

Fig. 16-9. A solid-state agc circuit which includes an agc delay amplifier stage.

tually becomes greater than the if bias. When the incoming signal is strongest, the rf amplifier is biased most heavily, and the signal is promptly reduced before it has a chance to overload any of the if amplifiers.

The solid-state agc circuit shown in Fig. 16-9 includes an agc delay stage which controls the agc voltage applied to the rf amplifier. The agc gate (Q403) functions as a variable resistance in series with diode CR402. The 1.36-volt emitter voltage for Q403 is established by diodes CR403 and CR404 which are forward biased by the 22.3-volt source through resistor R415. The base voltage for Q403 is obtained from the emitter of the video driver transistor through a voltage-divider network consisting of R413 and R414.

With no signal applied, agc gate Q403 is cut off and the −31 volts on the collector is developed through the rectification of the negative-going horizontal gating pulses by diode CR402. When a station signal is being received, a negative-going composite video signal is applied to the base of Q403. Each negative-going horizontal sync pulse within the composite video signal momentarily overcomes the no-signal reverse-bias condition. The tips of the horizontal sync pulses drive the base of Q403 less positive, causing the transistor to conduct. Collector current flows through CR402 and CR401 to charge C403, developing a positive voltage. The voltage developed by the charge on capacitor C403 is applied through resistor R408 to the base of agc output transistor Q402.

When a signal is being received, the collector of agc gate transistor Q403 becomes positive which

turns off the pnp transistor. However, the negative-going horizontal gating pulses fed to the collector through C402 and CR402 arrive at the same time as the horizontal sync pulses applied to the base. Therefore, Q403 is gated on by the horizontal gating pulses at the collector. This action is the same as that previously described for the agc keyer.

Under a no-signal condition, the base voltage for agc output transistor Q402 is derived through the voltage divider consisting of R411 and R412 connected between the 22.3-volt source and ground. The emitter voltage, with no signal applied, is derived through a voltage divider consisting of R405 and R406 which is also connected between the 22.3-volt source and ground. The voltages given in Fig. 16-9 are for a no-signal condition. When a strong signal is received, the output of the agc gate (Q403) is added to the base voltage of agc output transistor Q402. This results in a more positive emitter voltage which reduces the gain of the first video if amplifier stage through forward agc action. When the agc output voltage reaches approximately 6.5 volts, the gain of the first video if amplifier is at minimum. If a further reduction in signal strength is required, it is accomplished in the rf amplifier by the action of the agc delay amplifier, Q401.

The emitter of agc delay amplifier Q401 is connected directly to the emitter of the agc output stage. The base voltage for Q401 is supplied from a voltage-divider network consisting of R404, R403, and agc control R402. The 2.13 volts at the collector of Q401 is derived from a voltage divider on the tuner which is connected to a 11-volt dc regulated source. This 2.13-volt rf agc voltage maintains the rf amplifier at maximum gain during weak-signal conditions. Under strong-signal conditions, the agc output voltage at the emitter of Q402 increases to approximately 6.5 volts. As a result, agc delay amplifier Q401 conducts and the collector voltage rises above 2.13 volts, causing the rf gain to decrease after the first if amplifier has reached minimum gain.

QUESTIONS

1. What is the purpose of an automatic gain control circuit?

2. What is the difference between rectified agc and amplified agc?

3. How many signals are required in keyed agc? What are these signals?

4. What determines the grid voltage on the agc-keying tube during the burst of conduction?

5. What is the purpose of the resistor in the grid circuit of the keying tube?

6. What is the purpose of the clamper diode in the agc delay circuit?

7. What is the purpose of delayed agc?

EXERCISES

1. Draw the basic circuit of a keyed agc system. Show the input and output signals.

2. Explain the operation of the circuit in Exercise 1.

CHAPTER 17

Receiver Controls—Application and Adjustment

In the early days of television programming, most stations devoted many of their morning hours to the transmission of a test pattern which usually carried the station's call letters or distinguishing insignia. This enabled service technicians to check out and adjust the receiver during those hours. Today, however, most stations are "on the air" from early morning to late at night. What little test-pattern transmitting, if any, is usually done very late at night after regular programming is completed and only for the purpose of allowing station engineers to make tests and adjustments on the transmitter.

Since the use of some type of test pattern is by far the best way to make receiver adjustments and also check on such things as if response, ringing, video-amplifier response, etc., it is almost a necessity for the technician to have a test-pattern generator. Although at least one test-equipment manufacturer markets a service instrument which provides a standard test pattern, a color-bar generator with a crosshatch pattern will prove very useful. By using such a pattern, the technician can properly adjust the controls of the receiver for the best picture. The horizontal- and vertical-hold, centering, linearity, and focus controls can all be precisely adjusted while the test pattern is being displayed. We will now review these controls and show their effect on a standard test pattern.

Fig. 17-1 shows a typical test pattern displayed on the screen of a correctly adjusted television receiver. The pattern is similar to a more complex test chart (Fig. 17-2) developed by the television transmitter committee of the Electronic Industries Association (EIA). This standard resolution chart is used to test the performance of television transmitters as well as receivers. A study of its features will explain the use of the less-complicated test pattern in Fig. 17-1.

Fig. 17-2 shows the television resolution chart, with added explanatory letter symbols. The chart consists of a series of geometric forms and a number of tones ranging from black through gray to white. The gray scales determine whether all elements of the television system are preserving the correct ratios of light intensities (as video modulation) to accurately reproduce the televised scene at the picture tube. The circles are used in checking horizontal and vertical size and linearity in the picture.

The horizontal and vertical fan-shaped wedges are composed of lines which gradually narrow as they approach the center. An estimate of the resolution of the television system, including the receiver under test, can be determined by observing the point at which the lines can no longer be distinguished from one another. Numbers beside the horizontal and vertical wedges show the corresponding number of lines being reproduced when the individual lines of the fan can just be distinguished from each other. The vertical fans are used to determine the horizontal resolution of the system; conversely, the horizontal fans are used to determine the vertical performance of the system.

Tests for vertical and horizontal linearity and for such requirements as interlace are described in Fig. 17-2. We will discuss their application to the simplified pattern in Fig. 17-1 as we consider the effects on the reproduced pattern caused by maladjusted controls or malfunctioning circuits.

In our study of the circuits employed to produce the raster, the control of scanning by signal pulses, and the modulation and focusing of the cathode-ray beam, we covered the action of a number of variable controls. These controls can be classified into two distinct groups according to whether or not they are readily accessible to the viewer.

FRONT-PANEL OR VIEWER-OPERATED CONTROLS

The front-panel controls are usually on the outside front of the cabinet and can be operated by the viewer. They normally include the volume control and power switch, a channel selector for tuning to the desired channel, and a group of controls for adjusting the appearance of the picture. The picture control group lets the viewer adjust brightness and contrast of the picture. Fig. 17-3 shows the front-panel controls on various receivers.

PRESET CONTROLS

The preset (or service) controls may require adjustment during original installation or at infrequent intervals only and are not readily accessible

to the viewer. Their number and complexity differ somewhat among the various manufacturers. The preset controls most likely to require readjust-

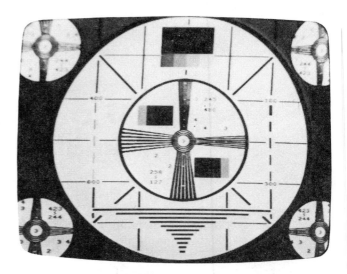

Fig. 17-1. Test pattern as it appears on a receiver that is operating normally.

ments can usually be reached without removing the chassis. They are adjusted by a knob or screwdriver. Fig. 17-4 shows the location of many of these controls in current-model receivers.

A. Vertical wedges (A) indicate horizontal line resolution.

B. Horizontal wedges (B) indicate vertical-line resolution.

C. Vertical-line boxes (C) check horizontal linearity.

D. Horizontal-line boxes (D) check vertical linearity.

E. Crosses at (E) are used to align the optical system of projection receivers.

F. Corner and center "bulls-eyes" (F) indicate spot shape and focus of cathode-ray tube.

G. Diagonal lines (G) check vertical interlace. Poor interlace causes lines to appear "jagged."

H. Shading bars (H) check visual fidelity of the system according to the number of steps ranging from light gray to black that can be distinguished.

I. Arrowheads (I) indicate boundaries of the usual area scanned by the camera as determined by the standard aspect ratio of 4 to 3.

J. Single resolution lines (J) indicate frequency of video-amplifier ringing.

K. Black horizontal bars (K) check low-frequency response or phase shift.

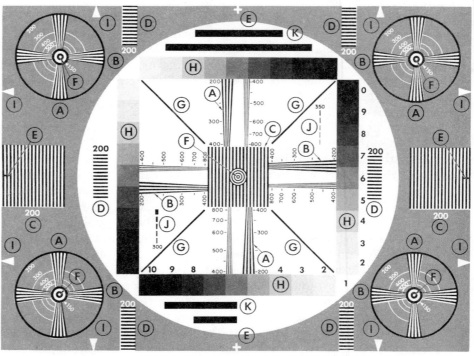

Fig. 17-2. A transmitter test chart.

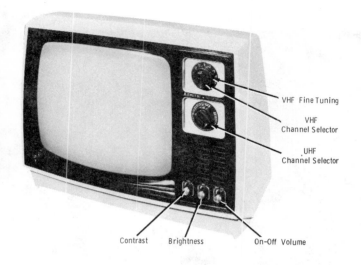

VHF Fine Tuning

VHF Channel Selector

UHF Channel Selector

Contrast Brightness On-Off Volume

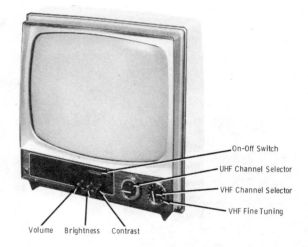

On-Off Switch

UHF Channel Selector

VHF Channel Selector

VHF Fine Tuning

Volume Brightness Contrast

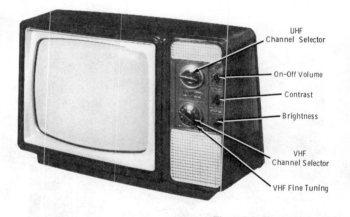

UHF Channel Selector

On-Off Volume

Contrast

Brightness

VHF Channel Selector

VHF Fine Tuning

Brightness

On-Off Volume

VHF Fine Tuning

VHF Channel Selector

UHF Channel Selector

Fig. 17-3. Typical examples of front-panel controls.

CLASSIFICATION ACCORDING TO FUNCTION

The controls of a television receiver can also be classified into four groups according to their functions.

1. Those which adjust the operating conditions of the cathode-ray picture tube.
   a. Adjustment of the deflection yoke—positions the raster correctly.
   b. Focus—for sharp definition. (Not used in most late-model black-and-white sets.)
   c. Adjustment of picture-tube operating voltages—for proper contrast between black level and highlight brightness.
   d. Scene brightness.

2. Those which establish the correct lock-in, or hold, of the horizontal- and vertical-scanning oscillators:
   a. Horizontal hold—sets the free-running frequency of the horizontal-scanning oscillator.
   b. Horizontal range or horizontal stabilizer—allows horizontal-hold control to be adjusted about its midpoint.
   c. Vertical hold—sets the free-running frequency of the vertical-scanning oscillator.
   d. Horizontal-oscillator frequency adjustment in afc systems.
   e. Horizontal-waveform adjustment in pulse-width systems.

3. Those which adjust the dimensions, shape, and position of the picture.

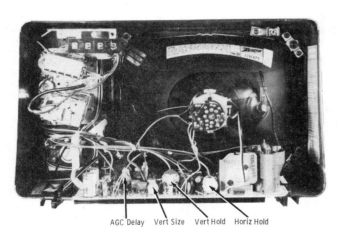

Fig. 17-4. Typical examples of preset controls.

a. Width control—adjusts horizontal size. (Not used on most late-model sets.)

b. Height control—adjusts vertical size.

c. Horizontal centering—moves the picture horizontally.

d. Vertical centering—moves the picture vertically.

e. Vertical linearity—controls the shape of the vertical-scanning wave.

f. Horizontal linearity—controls the shape of the horizontal-scanning wave. (Not used in most late-model sets.)

Deflection-Yoke Adjustment

Fig. 17-5 shows the test pattern when the deflection yoke is improperly positioned. The control of the electron beam by the deflection coils was explained in Chapter 2. If the lines of the raster are not horizontal and squared with the edge of the picture mask, the deflection yoke is incorrectly positioned. The yoke-adjustment lockscrew is loosened and the yoke is rotated until the raster lines are properly aligned with the edges of the picture mask. The yoke-adjustment lockscrew

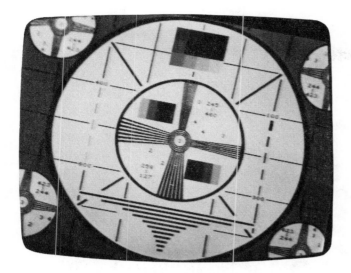

Fig. 17-5. Deflection yoke not properly oriented.

is then tightened. The position of the deflection yoke along the picture-tube neck will affect the deflection sensitivity (required scanning voltage for a given amount of deflection).

Focus-Control and Focus Connections

Fig. 17-6 illustrates the test pattern when the electron beam is out of focus. The image is not sharply defined, as it is in the normal picture in Fig. 17-1.

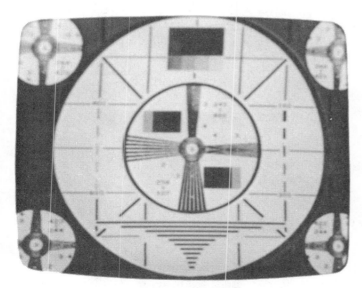

Fig. 17-6. Focus misadjusted.

If a focus control is employed, it should be re-adjusted. If the focus electrode is connected to a fixed-voltage source, the voltage should be changed to the appropriate value by moving the focus-lead

connector from the point where it is connected to a different voltage point.

Brightness Control

The brightness control normally is located on the front or side of modern receivers and can be adjusted by the viewer. The beam can be cut off with this control, and the tube will remain dark. Setting the brightness control too high causes a light, washed-out picture, as shown in Fig. 17-7. The shadows and lower-key halftones have disappeared.

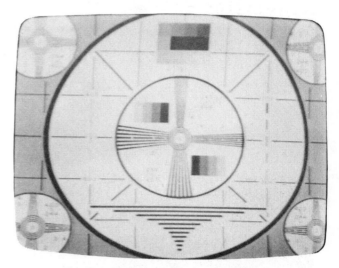

Fig. 17-7. Brightness control misadjusted (excessive brightness).

Hold Adjustments

The horizontal- and vertical-hold adjustments enable the free-running frequencies of the two receiver scanning systems to be adjusted for synchronism, or lock-in, with the transmitted sync pulse.

Horizontal Hold—Fig. 17-8 illustrates the effect on the test pattern when the picture is out of horizontal sync and the horizontal-hold control is adjusted to bring it back into sync. The diagonal bars become fewer and fewer until the picture snaps into synchronization. Proper hold-control setting is achieved when the picture does not go out of sync when the channel selector is switched off channel and then back on channel again.

Vertical Hold—Fig. 17-9 shows the test pattern when the vertical-hold control is misadjusted. The effects on the picture are similar to those discussed for the horizontal hold, except that the

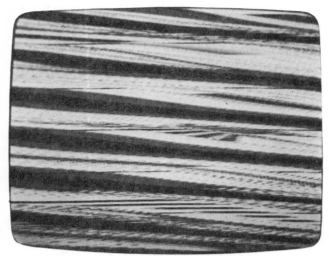

(A) Excessively out of sync.

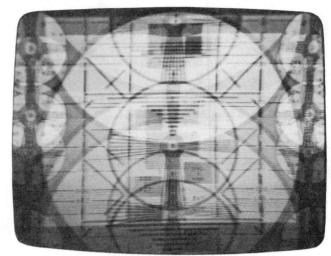

Fig. 17-9. Vertical-hold control misadjusted.

(B) Closer to sync.

Fig. 17-8. Horizontal-hold control misadjusted.

also affect the oscillator frequency. For this reason, the adjustment of the horizontal-oscillator frequency must be rechecked if the discriminator-phase control is changed. The service literature contains explicit instructions concerning the order in which these adjustments are to be made in a particular afc circuit.

The picture should remain horizontally synchronized when the horizontal hold control is turned to either extreme. To test for this condition, tune to a signal while the control is at midposition. Then turn the hold control to its extreme position in either direction. Next, remove the signal by detuning the receiver. When the receiver is retuned, the system should pull into sync. Make the same check at the other end of the control range. If the

image moves vertically instead of horizontally before lock-in occurs.

The vertical hold must be carefully adjusted. Otherwise, "pairing" of horizontal lines of alternate fields will occur. Pairing reduces the vertical definition of the picture.

Horizontal Oscillator Frequency Adjustment (AFC Systems)

Fig. 17-10 shows the test pattern when the horizontal-oscillator frequency is misadjusted. In the flywheel or afc system of horizontal-sync control, the free-running frequency of the oscillator is controlled by an iron core in an inductor. The horizontal-hold and the discriminator-phase controls

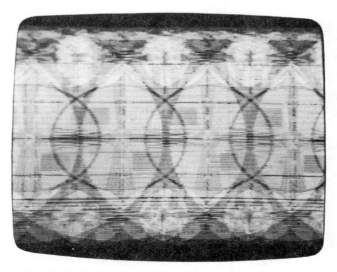

Fig. 17-10. Horizontal oscillator misadjusted (afc system).

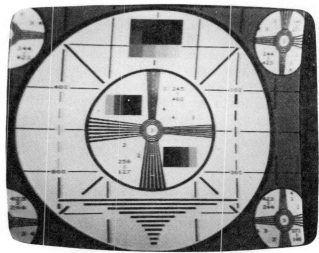

Fig. 17-11. Horizontal centering incorrect.

receiver does not pull into sync at both ends of the hold range, a readjustment of the horizontal-oscillator frequency should be made.

Centering Controls

The adjustments which center the picture are called the horizontal- and vertical-centering controls. In early receivers, one method of centering the picture was to vary the direct current through the deflection coils. This method is still used in some older color-television receivers. Another method of centering, used in early sets employing magnetic focusing, was to move the focus coil.

Almost all modern black-and-white receivers use two permanently magnetized rings on the neck of the picture tube to accomplish centering.

Fig. 17-12. Vertical centering incorrect.

The location of these rings is shown in Fig. 17-4. Fig. 17-11 shows the effect on the test pattern when the horizontal centering is misadjusted. An example of a picture which is not centered vertically is shown in Fig. 17-12.

Width Control

The width adjustment varies the energy applied to the horizontal-deflection coils. When the width control (or sleeve) is not properly adjusted, the picture is usually narrow. The width is usually controlled by regulating the output of the horizontal-output amplifier. Most modern receivers have no provision for controlling the width.

Height Control

The height control functions similarly to the width control, but in the vertical direction. The height control usually varies the amplitude of the sawtooth waveform applied to the vertical output stage. Fig. 17-13 shows the effect on the test pat-

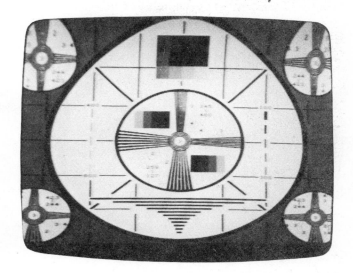

Fig. 17-13. Height control misadjusted.

tern when the height control is incorrectly adjusted. In most receivers, the height control has the greatest effect on the lower portion of the picture.

Vertical-Linearity Control

Fig. 17-14 shows the effect of a misadjusted vertical-linearity control on the test pattern. The vertical-linearity control is usually an adjustable bias resistor which adjusts the operating point of the vertical output stage. The vertical-linearity

Fig. 17-14. Vertical-linearity control misadjusted.

control normally affects the upper portion of the picture.

In most sets, the vertical-linearity adjustment is interdependent with the height control. Because of this interdependent action, one control may have to be adjusted if the setting of the other control is changed. In some sets, the control which normally functions as the height control is called the vertical-linearity control and vice versa. Therefore, it may be necessary to check the schematic in order to determine the function of a particular control.

QUESTIONS

1. What can be determined from the horizontal wedges of a test pattern? From the vertical wedges?

2. What are the purposes of the circles in the test pattern?

3. What adjustments are necessary when the picture is tilted, the lines are blurred, and the picture is off center?

4. Which controls are adjusted when the picture is distorted in the vertical direction?

5. Explain what is meant by front-panel or "viewer" controls. What are "preset" controls?

Appendix

Glossary

A

Absorption Trap. A parallel-tuned circuit coupled either inductively or capacitively to absorb and attenuate interfering signals.

Accompanying Audio (Sound) Channel. The rf carrier frequency which supplies the sound that accompanies the picture. Also known as co-channel sound frequency.

Active Lines. The lines which produce the picture, as distinguished from the lines occurring during blanking (horizontal- and vertical-retrace lines).

Adjacent Audio (Sound) Channel. The rf carrier frequency which carries the sound modulation associated with the next lower-frequency television channel.

Aluminized-Screen Picture Tube. A picture tube in which a thin layer of aluminum has been deposited on the back of the fluorescent surface. This layer improves the brilliance of the image and prevents ion-spot formation.

Amplitude Modulation (AM). A method by which an rf carrier is modulated. The instantaneous amplitude of the carrier is proportional to the instantaneous value of the modulating voltage.

Amplitude Separation. The method by which part of a voltage wave (particularly the sync pulse) is sorted from the rest of the wave by differences in amplitude.

Antenna Array. A system of antennas which are coupled together to produce a desired directional pattern or to increase signal pickup.

Aquadag Coating. A conductive coating on the surface of the glass envelope of picture tubes. The coating is formed by a colloidal solution of carbon particles. It is placed internally to collect secondary electrons emitted by the fluorescent screen. In most tubes it is also placed externally so that the tube can be used as a final capacitor for the high-voltage filter circuit.

Aspect Ratio. The ratio of picture width to picture height. The television standard is 4/3.

Asymmetrical-Sideband Transmission. (See Vestigial-Sideband Transmission.)

Audio Channel. (See Accompanying Audio Channel.)

Automatic Contrast Control. (See Automatic Gain Control.)

Automatic Gain Control (AGC). A method by which the overall amplification (gain) of a television receiver is automatically regulated, so that relatively constant output (contrast of the picture) is produced for varying input signals.

Automatic Phase Control. A method by which the frequency and phase of the horizontal-scanning oscillator are automatically held in step with the repetition rate and phase of the horizontal-sync pulses.

B

Background. In television, background is the average illumination of the scene and is represented by the dc component of the video signal. (Also see DC Video Component.)

Back Porch. The portion of the synchronizing signal (at blanking or black level) which follows the horizontal-sync pulse and precedes the start of the next horizontal active line. (Standard duration for this porch is 3.81 microseconds.)

Bandwidth. The difference in frequency between the highest and lowest frequencies involved. A television channel has a bandwidth of 6 MHz.

Beam. In a cathode-ray tube, the stream of electrons moving toward the screen.

Bidirectional. The shape of the reception pattern of an antenna that responds equally well to stations located 180° apart with respect to the antenna.

Blacker-Than-Black (Infrablack). That portion of the composite signal that is higher in amplitude than the black level (from 75% to 100% of maximum signal). This region is occupied by the synchronizing pulses.

Black Level. The amplitude of the composite signal at which the beam of the picture tube is extinguished (becomes black) to blank retrace of the beam. This level is established at 75% of the signal amplitude.

Blanking Pulses. The portion of the composite signal during which the beam of the picture tube is extinguished. (See Black Level.)

Blocking Oscillator. A type of relaxation oscillator for the generation of sawtooth waves. It employs an iron-core feedback transformer and a capacitor-resistor network with a long time constant in the input circuit.

Blooming. The phenomenon of the picture expanding when the brilliance is increased. Blooming is often caused by insufficient high voltage.

Booster. A separate rf amplifier connected between the antenna and the television receiver to amplify weak signals.

Brightness Control. This receiver control sets the operating point of the picture tube and determines the average brightness of the reproduced image.

C

Camera Tube. The device at the television transmitter that changes the light variations of the scene into electrical variations.

Cascode Circuit. A circuit which consists of a grounded-cathode triode driving a grounded-grid triode. Both triodes are connected in series, as far as plate current is concerned.

Cathode Follower. A circuit in which the input signal is applied between control grid and ground and the output is taken between cathode and ground. The plate is bypassed to ground. A cathode follower exhibits high input impedance and low output impedance.

Cathode Input. (See Grounded-Grid Amplifier.)

Centering Control. Controls the position of the raster on the picture-tube screen. (See Horizontal-Centering Control and Vertical-Centering Control.)

Channel. The band of frequencies assigned for the transmission of a television signal.

Characteristic Impedance. The ratio of the voltage to the current at every point on an rf transmission line. This ratio does not apply unless the line is properly terminated and there are no standing waves.

Clamping Circuit. A circuit which maintains the amplitude of a voltage wave at a predetermined dc level. (Also see DC Restoring.)

Clipping Circuit. A circuit that removes one or both extremities of an impressed voltage wave. This produces a flat-topped output.

Coaxial Cable (Coax). A high-frequency transmission cable consisting of a central inner conductor and a cylindrical outer conductor, the two being separated by an insulator.

Composite Signal. The transmitted television signal, which is composed of video modulation, blanking pedestals, and synchronizing pulses. Blanking pedestals occur at 75% of maximum signal, and the sync pulses occupy the remaining 25%.

Contrast. The range of light-to-dark values of the image, proportional to the voltage swing at the picture-tube input.

Contrast Control. Controls the voltage swing to the picture-tube input, so that the most acceptable image contrast is produced.

D

Damper. A diode that momentarily short-circuits the stored energy in the horizontal-sweep system and thus prevents transient oscillations.

DC Restoring. The combining of the dc component of the video signal (lost in capacitance-coupled amplifiers) with the ac component, so that the average light value of the reproduced picture is re-established.

DC Video Component. The part of the video signal caused by the average steady background illumination of the scene being televised.

Definition. The sharpness of the fine detail in the reproduced picture.

Deflection. The process whereby the electron beam of a picture tube is deviated from its axial path to produce the raster.

Deflection Yoke. An assembly consisting of the horizontal- and vertical-deflection coils used in magnetic deflection systems.

Detail. The least perceptible elements or areas of an image which can be recognized as being different from one another.

Diathermy. An interfering signal caused by certain types of medical apparatus. The interference usually produces a "herringbone" pattern on the picture.

Differentiating Circuit. A combination of circuit elements which produces an output pulse proportional to the rate of change of the input signal.

Dipole. An antenna whose length is approximately one-half the electromagnetic wavelength to which it is resonant. The antenna is usually divided in the middle for connection to the transmission line.

Director. An antenna element placed in front of a dipole element (toward the transmitter) to increase the pickup and obtain a directional pattern.

Discharge Tube. A tube which discharges a wave-shaping capacitor. The tube is biased to cutoff until triggered by a positive pulse.

Discriminator. An fm-detector circuit using a pair of diodes. The audio output is proportional to the frequency deviation.

Drive Control. (See Horizontal-Drive Control.)

Driven Element. The antenna element that is connected to the transmission line.

Dynode. In a photomultiplier tube, an additional electrode which produces secondary emission when bombarded by electrons.

E

Echo. A delayed signal. This may refer either to a reflected television carrier (ghost signal) or to effects in the video amplifier.

Electromagnetic-Deflection Coil. A circular coil placed near the neck of a picture tube, producing a magnetic field which deflects the electron beam.

Electron. The elementary electrical charge of negative polarity (1.6×10^{-19} coulombs).

Electron Focus. The reduction in size of the electron beam in a picture tube by variation of an electrostatic field.

Electron Gun. An arrangement of electrodes which produces and controls a small beam of electrons in a cathode-ray tube.

Electron Multiplier. An electron tube in which a number of electrodes are arranged in cascade. Each electrode delivers more electrons to the next electrode than it receives. The increase is due to secondary emission.

Electron Scanning. The deflection of an electron beam to form a regular pattern or raster.

Electrostatic Field. A strain in space which exerts a force on an electrical charge (electron) within its region of influence.

Electrostatic Scanning. The deflection of an electron beam by means of an electrostatic field.

Equalizing Pulses. A series of pulses (usually six) which occur at twice line frequency and which precede and follow the serrated vertical-synchronizing pulse. These pulses cause vertical retrace at the correct instant for proper interlace.

F

Field. One complete scanning operation of the picture from top to bottom, including vertical retrace. This scanning takes 1/60th of a second and occurs twice per frame.

Field-Effect Transistor. A transistor with a main conduction path through a bar of n-type or p-type silicon, with control through a field on each side of the bar. This type of transistor is employed in some of the latest solid-state receivers.

Field Frequency. The number of fields transmitted per second. Sixty fields per second is standard.

Field-Frequency Synchronizing Pulse. (See Vertical Blanking.)

Field Period. The time required for one field to be transmitted. This period is the reciprocal of the field frequency, or 1/60th of a second.

Fluorescent Screen. The phosphor coating on the inside of the picture-tube faceplate, which emits light under the impact of the electron beam.

Flyback. The return of the electron beam after tracing a horizontal line. (See Retrace.)

Flywheel Sync. An early method of horizontal synchronization in which scanning is controlled by the average timing of the sync pulses rather than by each individual pulse, as in triggered sync.

Focusing Control. Brings the electron beam to the smallest possible spot on the fluorescent screen.

Folded Dipole. An antenna consisting of two half-wave dipoles parallel to each other and having their extremities connected. One of the dipoles is open at its center point for connection to the transmission line.

Forward Bias. A bias voltage applied to a solid-state device (transistor or diode) in such a direction as to cause current conduction through the device. Opposite of reverse bias.

Frame. The complete single picture contained in an image. A frame consists of two fields and has a repetition rate of 30 frames per second.

Frame Frequency. The number of times per second the picture is completely scanned—the standard is 30 frames per second.

Front Porch. The portion of the synchronizing signal (at blanking or black level) which precedes the horizontal-sync pulse and which occurs at the end of each active horizontal line (1.27 microseconds is standard duration for the front porch).

Front-to-Back Ratio. The antenna sensitivity ratio of signals arriving from the front (desired direction) to the signals arriving from the back (180° from the front).

G

Ghost Image. A second, or echo, image superimposed on the picture because of a reflected carrier wave.

Grid Limiting. Grid-current bias derived from the signal, through a large series grid resistor, which cuts off the plate current and thus levels the output wave for all input signals above a critical value.

Grounded-Grid Amplifier. A circuit in which the input signal is applied between a grounded grid and the cathode. The grid thus acts as a shield between the input circuit (cathode-to-ground) and the output circuit (plate-to-ground).

H

Halation. A halo surrounding a point of high brilliance on the fluorescent screen. The halo may be due to light scattering in the phosphor or to multiple reflections between the front and back surfaces of the glass.

Half-Wave Dipole or Doublet. (See Dipole.)

Height Control. Controls the amplitude of the vertical-sweep pulses and hence the height of the picture.

Hold Control. Controls the scanning-oscillator phase and frequency (horizontal or vertical), so that it is synchronized with the transmitted sync signal.

Horizontal Blanking. The process of cutting off the electron beam during retrace between successive active horizontal lines.

Horizontal-Blanking Pulse. The rectangular pedestal of the composite television signal, which occurs between active horizontal lines. This cuts off the beam current of the picture tube during horizontal retrace.

Horizontal-Centering Control. Moves the image horizontally on the picture-tube screen.

Horizontal Directivity. In a receiving antenna, the reception characteristic of the antenna in the horizontal plane.

Horizontal-Drive Control. Controls the ratio of the negative pulse amplitude to the linear portion of the scanning-voltage wave. Used in early tube-type television receivers.

Horizontal Flyback. (See Retrace.)

Horizontal-Hold Control. (See Hold Control.)

Horizontal Polarization. A transmitted signal is horizontally polarized when its electrostatic component is parallel to the earth's surface.

Horizontal Repetition Rate. The number of horizontal lines per second (15,750).

Horizontal Resolution. The number of horizontal picture elements which can be distinguished from each other in a single active line.

Horizontal Retrace. (See Retrace.)

Horizontal-Scanning Frequency. (See Horizontal Repetition Rate.)

Horizontal Stabilizer. A control used in some television receivers for adjusting the range of the Horizontal-Hold control.

Horizontal-Sync Discriminator. In the flywheel method of synchronization, a circuit which compares the phase of the horizontal-sync pulses with that of the horizontal-scanning oscillator. The output of the discriminator controls the frequency and phase of the oscillator by means of a reactance tube.

Horizontal-Sync Pulse. The rectangular pulse which occurs, above blanking level, between each active horizontal line. Because of these pulses, horizontal scanning of the receiver coincides with that of the transmitter.

Hum Bar. A dark band extending across the picture, caused by excessive 60-Hz hum (or harmonics thereof) in the input signal of the picture tube.

I

Iconoscope. A camera tube in which an optical image on a mosaic of photosensitive elements causes charges which are proportional to the image intensity at each point. A scanning beam releases these charges as a video signal.

Image Interference. A spurious response in a television receiver due to signals at a frequency twice the intermediate frequency. Usually due to an fm carrier or another television carrier.

Image Orthicon. A camera tube combining the orthicon principle with that of the electron multiplier. This tube is extremely sensitive to light.

Implode. The term applied to the bursting of a picture tube. Because of the high vacuum in the tube, the glass fragments move inward with terrific force.

Infrablack. (See Blacker-than-Black.)

Integrated Circuit. Used in many of the later-model television receivers. An interconnected array of circuit components fabricated on and in a single chip of semiconductor material by diffusion, etching, etc., and capable of performing a complete circuit function.

Integrating Circuit. A combination of circuit elements which produces an output potential proportional to the stored-up value of a number of pulses of input signal.

Intercarrier Sound. A method by which the 4.5-MHz difference between the video and sound carriers is used as an intermediate frequency for the sound signal.

Interlace Scanning. A method by which an image is scanned in successive fields. Each field contains only part of the horizontal line structure, the fields being arranged so that successive fields supply the lines missed by the preceding fields. The lines of the second field fall exactly between the lines of the first. This system overcomes the flicker which would occur if areas on the screen were scanned vertically at a rate of only 30 Hz.

Ion. A charged atom. In the picture tube, ions produce a dark spot at the center of the fluorescent screen unless they are blocked out of the electron beam.

Ion Spot. A dark spot on the fluorescent screen of a picture tube, caused by ion bombardment.

K

Keyed AGC. An agc circuit in which a pulse from the high-voltage transformer is applied to the plate or collector, so that the stage can conduct during horizontal sync time only.

Keystone. A trapezoid. A television picture with one side shorter than the other has keystone distortion.

Keystone Distortion. (See Keystone.)

L

Limiter. A stage preceding a discriminator-type fm detector. All input signals above a predetermined amplitude drive this stage to cutoff and thus limit the value of the output signal. The limiter removes amplitude modulation.

Line Frequency. (See Horizontal Repetition Rate.)

Line-Frequency Blanking Pulse. (See Horizontal-Blanking Pulse.)

Linearity. The distribution of picture elements over the image field, as determined by the shape of the horizontal- and vertical-scanning waves. In a linear picture, the elements are uniformly and correctly distributed. If the scanning motion is nonlinear, the picture will be distorted.

Linearity Control. Corrects distortion of the sawtooth voltage or current waves used for deflection.

M

Magnetic Deflection. The method by which a magnetic field, produced by a coil external to the picture tube, moves the electron beam. A linear sawtooth motion is produced when the current through the coil has a linear sawtooth form.

Magnetic Field. The space near a permanent magnet or electromagnet in which a force is exerted upon an electron.

Marker Pip. A frequency index mark used with a sweep generator during alignment of tv receivers. The marker pip is produced by a fixed-frequency oscillator which is coupled to the output of the sweep generator.

Monitor. A picture tube and its associated circuits which are usually connected directly to the video system of the transmitter and which permit the transmitted picture to be viewed.

Monoscope. A cathode-ray tube which produces a video signal for testing purposes. An internal test pattern is employed.

Mosaic. A photosensitive surface used in some camera tubes and consisting of an insulating surface covered with numerous photosensitive "islands." A plate behind the insulator collects the charges when the mosaic is scanned by an electron beam.

Multipath Reception. The reception of a direct wave from the television transmitter, accompanied by one or more reflected and delayed waves. (See Ghost Image.)

Multiplier. (See Electron Multiplier.)

Multivibrator. A relaxation oscillator which produces sawtooth waves. Two active circuit elements (tubes or transistors) are used; the output of each is coupled to the input of the other through RC networks, which determine the period of oscillation.

N

Negative Transmission. The polarity of modulation of the standard television signal. The sync pulses and signals corresponding to black drive the carrier toward maximum amplitude; signals of highest brilliance (white) drive the carrier toward zero amplitude.

Neutrode Circuit. A neutralized triode amplifier in which a capacitor feeds back voltage to the control grid from the low side of the plate coil.

Noise. Interference that causes a salt-and-pepper or "snowy" pattern on the picture. It is due to atmospherics, tube-fluctuation effects, or man-made interferences.

Nonlinearity. The crowding of picture elements at the sides, top, or bottom of the picture because scanning is not linear. (See Linearity.)

O

Odd-Line Interlace. A double interlace system in whi there is an odd number of lines per frame. Therefore, ea field contains a half-line.

P

Pairing. Improper interlace, in which the horizontal lines of alternate fields do not fall exactly between those of the preceding field. When pairing is most pronounced, the lines of alternate fields fall on one another and result in separated lines with half the possible vertical resolution.

Parasitic Element. An antenna element not coupled directly to the transmission line.

Peaking Coil. In a video amplifier, an inductance which resonates with the circuit capacitance near the upper limit of the passband. Thus, high-frequency loss of gain is compensated for, and the amplifier phase shift is corrected.

Pedestal. The level of the video signal at which blanking of the picture-tube beam occurs. (See Blanking Pulses and Black Level.)

Persistence of Vision. A characteristic of the eye and brain whereby the sensation of an image remains after the light causing it has vanished. This effect lasts for approximately an eighth of a second and makes television as well as motion pictures possible.

Phase. A point in the cycle of an alternating current or voltage with respect to zero time.

Phase Distortion. A condition of different phase delays for different video frequencies. This action causes distortion of the peak values of the video signal and results in poor contrast and resolution.

Photoelectric. The phenomenon whereby electrons are emitted from certain substances due to absorption of light.

Pickup Tube. Same as Camera Tube.

Picture Element. The smallest of picture areas that can be distinguished from each other.

Picture Tube. In television receivers, a cathode-ray tube which translates the video signal into a picture.

Polarization. The direction (either horizontal or vertical) of the electric field of a radiated wave. The magnetic field is perpendicular to the electric field.

Pre-emphasis. The practice in which the high frequencies of the audio spectrum are amplified more than the low frequencies. Employed in fm transmission of the television sound channel.

Printed Circuit. A circuit using conductive strips rather than wire. The strips are printed, etched, etc., onto an insulating board. The circuit may also include components which have been formed in a similar manner. Most late-model television receivers employ printed circuits.

Q

.h
h

.erit of a capacitor, an inductor, or a
s equal to the ratio of the reactance to

Having properties similar to light waves.
pagation of waves in the television spectrum
"quasi-optical," i.e., cut off by the horizon.

e Sideband Transmission. Same as Vestigial-
Transmission.

R

Detector. An fm detector which discriminates against
.tude modulation. A pair of diodes is so connected that
audio output is proportional to the ratio of the fm volt-
.s applied to the two diodes.

C Circuit. A time-determining network of resistors and
capacitors in which the time constant is defined as the
product of the resistance times the capacitance.

Reactance-Tube Circuit. A circuit in which a high transconductance tube is so connected that it appears as a reactance (inductive or capacitive) to a circuit across which it
is connected. The value of the reactance can be controlled
by changes in the dc grid bias. Used for afc.

Reflection. The reflected carrier wave and the ghost image
on the picture, caused by the reflected carrier.

Reflector. An element placed behind the pickup element of
a receiving antenna to intensify the receiving signal and
improve the shape of the directional pattern.

Registry. The superposition of one image on another, as in
the formation of an interlaced raster.

Reinserter. (See DC Restoring.)

Relaxation Oscillator. An oscillator which generates periodic waves in which a sudden excursion of plate current
from cutoff to saturation and back is followed by a relatively long period of quiescence, or relaxation.

Resolution. In television, the ability of a receiver to reproduce picture detail. It is usually expressed as the number
of lines which can be seen on a reproduced test chart.

Resolution Pattern. (See Television Test Pattern.)

Retrace. The return of the electron beam after a horizontal or vertical scan.

Retrace Time. The time which elapses during retrace. The
time is approximately 7 microseconds for the horizontal
retrace and 500 to 750 microseconds for the vertical retrace.

Reverse Bias. A bias voltage applied to a solid-state device
(transistor or diode) in such a direction as to prevent or
cut off conduction through the device.

Rhombic Antenna. A diamond-shaped arrangement of conductors, all of the same length (rhombus) and joined at
three corners with a transmission line connected at the
open corner. The length of the conductors must be more
than a wavelength. Used for fringe-area or low-signal reception.

S

Sawtooth. The waveform employed in television scanning.

Scanning. The process by which the light values of the picture elements which constitute the entire scene being televised are successively analyzed according to a predetermined method.

Scanning Generator. A circuit which produces the sawtooth
wave of voltage or current required for proper beam deflection in the camera tube or picture tube.

Scanning Line. A single active horizontal line of the picture.

Scanning Spot. The cross section of the electron beam at
the fluorescent screen of the picture tube.

Screen Persistence. The property by which the fluorescent
screen continues to radiate light for a short time after the
electron beam has passed.

Secondary Electron. An electron which is emitted from a
metal surface under the bombardment of another electron
called the primary electron.

Secondary Emission. The phenomenon by which secondary
electrons are produced. (See Secondary Electron.)

Series Peaking. An inductance in series with the video-
amplifier plate circuit, which compensates for loss of high-
frequency gain and corrects high-frequency phase shift.

Serrated Pulses. In the long vertical-sync pulse, "notches"
which keep the horizontal oscillator synchronized during
vertical retrace.

Shunt Peaking. An inductance in parallel with a video-
amplifier load. It compensates for the high-frequency loss
due to the shunt capacitance and corrects the high-frequency phase shift.

Single-Ended. Input circuits of television receivers in which
one side of the transmission line is connected to the chassis
or ground.

Single Sideband. A method of rf transmission in which the
carrier and only one sideband are radiated.

Smear Ghost. A spurious image caused by multiple reflections or by phase shift in the video amplifier.

Snow. Television slang for the effect of random noise in the
reproduced picture. (See Noise.)

Spurious Signal. Refers either to the effect of reflections of
the carrier wave or to undesirable shading signals in the
camera tube.

Stacked Arrays. Antenna systems in which two or more
antenna arrangements are positioned above each other at
a critical spacing and connected by transmission line.
Stacking increases the pickup and improves the directional
pattern.

Staggered Circuits. Interstage coupling circuits of a video if amplifier are staggered when they are tuned to different frequencies. Staggered tuning permits broadband response.

Surge Impedance. Same as Characteristic Impedance.

Sweep. Refers to the displacement of an electron beam from its axis of origin. (Also see Scanning.)

Sweep Voltage. The voltage applied to the deflection plates or coils of a cathode-ray tube.

Sync. Abbreviation for synchronizing.

Sync Clipper. A circuit which is biased so that the sync signals are removed from the composite video signal.

Sync Inverter. A circuit which produces a 180° phase shift of the sync pulses. Thus, necessary signal polarity for control of the scanning oscillator is provided.

Sync Leveler. (See Sync Limiter.)

Sync Limiter. A circuit which produces sync pulses of uniform height.

Sync Separator. (See Sync Clipper.)

Synchroguide. A type of control circuit for horizontal scanning in which the sync signal, oscillator voltage pulse, and scanning voltages are compared and kept in synchronism.

Synchronizing Pulses. The portions of the transmitted signal which control horizontal and vertical scanning of the receiver. (See Horizontal- and Vertical-Sync Pulses.)

T

Tearing. A fault of the synchronizing system in which groups of lines are displaced; this causes the appearance of a torn picture.

Televise. To convert a scene or image field into a television signal.

Television Test Pattern. A televised chart with geometric patterns by which the service technician or set user can determine the operating condition of the receiver.

Time Constant. The time required for the voltage or current of a circuit to rise to 63% of its final value or to fall to 37% of its initial value.

Time Delay. The elapsed time between the start of a modulation wave or pulse at the transmitter and the start of its reproduction on the picture-tube screen.

Transient Response. The manner in which a circuit responds to sudden changes in applied potential.

Transmission Line. A two-conductor circuit with uniformly distributed electrical constants, used for transmitting radio-frequency signals.

Triggering. Starting of an action in a circuit, which then continues to function for a predetermined time under its own control.

U

Ultrahigh Frequencies (UHF). The portion of the electromagnetic-radiation (radio-wave) spectrum from 300 to 3000 MHz.

V

Varactor Diode. A reverse-biased diode which can be utilized as a variable capacitor by changing the voltage across the pn junction.

Varactor Tuner. A tuner in which channel selection and tuning is accomplished by applying a common control voltage to varactor diodes in the tuned circuits.

Vertical Blanking. The blanking signals at the end of each vertical field. They blank out the picture tube during vertical retrace.

Vertical-Blanking Pulse. (See Vertical Blanking.)

Vertical-Centering Control. Moves the picture vertically on the picture-tube screen.

Vertical Directivity. In a receiving antenna, it is the reception characteristic of the antenna in the vertical plane.

Vertical-Hold Control. (See Hold Control.)

Vertical Oscillator. The sawtooth-scanning generator which furnishes the required voltage or current wave for vertical scanning.

Vertical Polarization. A transmitted signal is vertically polarized when its electrostatic component is at right angles to the earth's surface.

Vertical Resolution. The number of picture elements which can be resolved vertically.

Vertical Retrace. The return path of the electron beam (blanked out) from bottom to top at the end of each field.

Vertical Scanning. The vertical movement of the beam on the picture tube.

Vertical-Scanning Generator. (See Vertical Oscillator.)

Vertical-Sync Pulses. Between each field, a series of six pulses which synchronizes the vertical-scanning oscillator.

Very High Frequencies (VHF). The portion of the electromagnetic-radiation (radio-wave) spectrum from 30 to 300 MHz.

Vestigial-Sideband Transmission. The standard system of video modulation. The carrier is modulated by a complete upper sideband and a vestige (1.25 MHz) of the lower sideband.

Video. (Latin for "I See.") The frequencies employed for modulation of a television carrier.

Vidicon. A camera tube which has a photoconductive mosaic and which operates similarly to the Image Orthicon.

W

Width. The horizontal dimension of the picture or the time duration of a pulse (pulse width.)

Width Control. Controls the horizontal dimension of the picture so that it fills the picture-tube screen.

Y

Yagi Array. An arrangement of dipole antenna elements employed for television reception. One element acts as the antenna; and the others act as parasitic elements (directors and/or a reflector), so that gain and the directional reception pattern are improved.

Answers to Questions

CHAPTER 1

1. Cathode, control grid, focus anode (anode 1), and acceleration anode (anode 2).
2. An indirectly heated cathode.
3. The quantity of electrons in the beam, which is governed by the control grid.
4. Focusing anode.
5. Deflection sensitivity.
6. The velocity of the beam and the strength of the deflecting field.
7. Raster.

CHAPTER 2

1. Right angle.
2. When the electron stream runs parallel to the lines of force in the magnetic field.
3. A focusing anode.
4. (a) Automatically.
   (b) By manually changing the potential applied to the focus grid.
5. A deflection *yoke* which consists of two sets of coils placed around the neck of the tube. One set of coils is for horizontal deflection and the other set is for vertical deflection.
6. A sawtooth current.
7. By rotating two magnetized rings located directly behind the deflection yoke on the neck of the picture tube.
8. The ions must be removed. If the ions are allowed to strike the phosphor screen, a brownish, burned area will result because they are much heavier than electrons.

CHAPTER 3

1. It translates the scene to be transmitted into equivalent electrical pulses.
2. A semitransparent photocathode.
3. (2) Monoscope.
4. 30 frames per second and 525 horizontal lines per frame.
5. Vertical—60 Hz.
   Horizontal—15,750 Hz.
6. Interlaced scanning.

7. The output polarity of the vidicon is positive.
8. The target in the image orthicon is the charged surface scanned by the electron beam.
9. The image is focused on one side of the photoconductive material in the vidicon and the other side is scanned by the electron beam.
10. The video signal must contain the blanking pulse and the synchronizing pulses in addition to the picture information.

CHAPTER 4

1. Higher current is required and good regulation is necessary.
2. No bulky power transformer is required.
3. Without the power transformer, there is a shock hazard because one side of the power line is connected to the chassis.
4. Voltage-divider networks are used to reduce the voltage to the desired value.
5. Power supplies used in solid-state receivers must be well regulated.
6. Due to the relatively low voltages used in solid-state television receivers, a small voltage change could seriously affect the operation of a transistor.
7. An electronic filter usually has a power transistor connected in series with the regulated portion of the load. The collector-to-emitter resistance of the regulator transistor is determined by the bias voltage on the base. If the power-supply voltage tends to rise, a reverse-bias voltage is applied to the base of the regulator transistor, and its collector-to-emitter resistance increases. The increased series resistance of the regulator transistor will oppose the change that initiated the action. A drop in the power-supply voltage will cause an opposite action.
8. Any line ripple in the power supply will show up as hum modulation in the picture and sound. Also, adequate decoupling must be provided for the sweep oscillators.

CHAPTER 5

1. In a charged capacitor, one plate contains more free electrons than the other plate. When a capacitor is dis-

charged, both plates contain the same number of free electrons.

2. The voltage across the capacitor increases, and the voltage across the resistor decreases.

3. These curves are nonlinear. They are exponential curves.

4. At the first instant, the voltages across the resistor is equal to the voltage across the fully charged capacitor, then gradually decrease as time progresses.

5. The time required to charge a capacitor to 63.2% of the applied voltage. T (in seconds) = R (in ohms) × C (in farads).

6. Integrator voltage: differentiator voltage.

7. The voltage across the capacitor decreases, and the voltage across the resistor approaches that of the input voltage.

CHAPTER 6

1. Neon and thyratron.

2. By use of a capacitor connected from the output of the second stage to the input of the first stage.

3. The discharge of the coupling capacitors.

4. The time constant of the RC circuit of one stage is made greater than that of the other stage.

5. A common cathode resistor is employed, and a coupling capacitor is not used from the plate of the second stage to the grid of the first stage.

6. The sawtooth-forming capacitor is allowed to discharge through the tube.

7. The current through the primary winding sets up a magnetic field and a secondary voltage is induced across the grid winding, causing a positive potential to appear at the grid.

8. The magnetic field in the transformer collapses, and the voltage induced in the secondary is such that it causes the grid to go more negative.

9. (a) Symmetrical square waves.
   (b) A short rectangular pulse followed by a long gap.
   (c) A short sine-wave followed by a long interval of relaxed oscillation.

10. Blocking oscillator and multivibrator.

CHAPTER 7

1. 1/60th of a second.

2. 15,750.

3. Triggers the vertical oscillator, blanks out the screen during retrace, keeps the horizontal oscillator in step during retrace, and provides for proper interlace.

4. (c) Equalizing pulses.

5. A positive pulse controls the blocking oscillator. A negative pulse controls the cathode-coupled multivibrator.

6. A sawtooth current is passed through the coils. The voltage waveform applied is a combination sawtooth and square wave.

CHAPTER 8

1. To stop, or damp out, oscillations in the horizontal-sweep circuit and help produce the required linear sawtooth current through the horizontal deflection coils.

2. Inductive.

3. The first half.

4. Anywhere from 5000 to 30,000 volts.

5. A high-voltage pulse. During retrace time.

6. To produce a dc control voltage for varying the frequency of the horizontal oscillator in a direction that will assure synchronization with the incoming sync-pulse rate.

7. To permit the horizontal-hold control to be set so that it operates about its midpoint.

8. To nullify any shock-excited oscillations following vertical retrace.

9. An integrator network rejects the horizontal-sync pulses. A differentiating network accepts the horizontal sync pulses.

CHAPTER 9

1. 30 Hz to over 4,000,000 Hz.

2. 4:3.

3. The brightness of the spot decreases.

4. 6 MHz.

5. Vestigial-sideband modulation.

6. At 1.25 MHz above the lower limit of the channel. The limit of the upper sideband is 4 MHz above the picture carrier.

7. 75%. Not more than 15% or less than 10% of maximum carrier amplitude.

8. Maximum.

CHAPTER 10

1. (a) At the video-detector input.
   (b) At any of the video-amplifier stages.
   (c) At the point of dc restoration.

2. To remove the sync signals from the composite video signal.

3. Positive.

4. The gain is high enough that the output signal does not need amplification.

5. A sync-phase inverter.

6. Across a resistor. Across a capacitor.

7. They differ in time duration.

8. Smaller. Total time-constant calculation is the same as for resistors in parallel:

$$\frac{1}{T} = \frac{1}{T1} + \frac{1}{T2} + \frac{1}{T3}$$

CHAPTER 11

1. Horizontally polarized. Horizontal polarization is the standard for the United States. However, the FCC has recently approved circular polarization for some stations.

2. The matching of the characteristic impedance of the transmission line to the impedance of the receiver and the antenna.

3. When the antenna is rotated, the ghost images due to line reflections will remain stationary, but the ones due to signal reflections will vary.

4. When the dipole is horizontal and at right angles to the direction of the transmitting antenna.

5. 72 ohms; 300 ohms.
6. The director is a parasitic element that is placed in front of the receiving antenna and is used to direct the signal to the antenna. The reflector is a parasitic element that reflects to the antenna a portion of the signal that passes by the antenna. The director is located in front of the receiving antenna and the reflector is located behind the receiving antenna. The reflector is longer than the director.
7. A stacked array consists of two or more identical antennas, one mounted above the other in the same vertical plane and spaced the proper distance apart.
   (a) Additional gain is obtained.
   (b) Some vertical directivity is contributed by the mutual interaction of the antennas.

CHAPTER 12

1. Rf amplifier, mixer, local oscillator.
2. The rf amplifier.
3. By a capacitor connected from the low side of the plate coil to the grid of the tube.
4. The input, rf, mixer, and oscillator coils.
5. The features of the turret and switch-type tuners are combined in the disc-type tuner.
6. Tuning is accomplished by applying a control voltage to a varactor diode which changes its characteristic capacitance.
7. The function of the aft circuit is to adjust the local oscillator for optimum tuning.
8. Forward agc is used in a transistor rf amplifier with relatively high resistance in the collector circuit. When the forward bias is increased, the additional collector current causes the transistor to operate in its saturation region and the gain of the stage is reduced.
9. It is difficult to amplify uhf frequencies efficiently. Additional gain for the converted uhf signal is provided by the rf amplifier and mixer stages of the vhf tuner.

CHAPTER 13

1. Fine detail of the picture is lost.
2. (a) Overcoupled transformer with shunt-resistance loading.
   (b) Stagger-tuned circuits.
3. The Q and gain are lowered and the response curve flattens out.
4. Instead of all stages tuned to the same frequency, each stage is tuned to a different frequency about the center frequency.
5. The response of the video if circuits must be reduced at the position which is occupied by the sound carrier of the channel being received, the video carrier of the next higher adjacent channel, and the sound carrier of the next lower adjacent channel.
6. Positive.
7. Negative going; positive going.
8. To prevent the if amplifier from oscillating due to feedback.
9. An absorption trap is a parallel-tuned circuit inductively coupled to an if stage to attenuate the response at a given frequency.

10. Because the capacitive-reactance of the detector diode and its associated circuit is low at the high video frequencies.

CHAPTER 14

1. 4.5 MHz.
2. 25 kHz.
3. The video-detector output or the video-amplifier output.
4. Ahead of the 4.5-MHz sound trap.
5. The ratio detector.
6. Series.
7. Because the ratio detector is not sensitive to amplitude modulation.
8. Limiter, audio detector, and 1st audio amplifier.
9. Limiting in the locked-oscillator detector circuit depends on the damping of strong signals in the grid circuit. A strong input signal causes the control grid to draw considerable current, which loads down the tuned circuit connected to the grid. The oscillation is suppressed by the grid loading.
10. An integrated-circuit chip which functions as the sound if amplifier, audio detector, and audio amplifier.

CHAPTER 15

1. The result is improper contrast of larger areas in relation to the smaller objects in the picture.
2. By using a large electrolytic bypass capacitor.
3. Fine lines and small picture elements will be blurred or missing.
4. The shunting effect of the stray capacitance of the various elements to ground. The loss is minimized by shunt- or series-peaking coils. A combination of the two is often used.
5. The brightness control adjusts the bias between the control grid and the cathode of the picture tube.
6. (a) The contrast control can be connected to change the bias of the stage.
   (b) If the control-grid resistor is connected to the cathode end of the contrast control, which is in the cathode circuit of the video-amplifier stage, the effective B+ voltage on the tube is varied.
   (c) The contrast control can be connected in parallel with the plate-load or collector-load resistor to tap off a portion of the video signal.
7. The setting of the brightness control.

CHAPTER 16

1. To minimize the effect of changes in signal strength at the receiving antenna.
2. With amplified agc, the rectified agc voltage is amplified before being applied to the controlled stage.
3. Two; a composite video signal which contains horizontal-sync pulses, and pulses from a winding on the horizontal-output transformer.
4. The level of the sync pulses in the composite video signal applied to the grid.
5. It limits the amount of grid current that can be drawn. The impedance of the resistor also isolates the input capacitance of the keying tube from the video circuit

and, hence, the loading effect on the video circuit is greatly reduced.

6. When the rf agc voltage tends to become positive, the clamper diode conducts and effectively shorts the agc line to ground.

7. To prevent rf amplifier agc action on weak signals and, thereby, to permit the rf amplifier to operate at maximum gain.

CHAPTER 17

1. The horizontal wedges are used to determine the vertical resolution of the receiver. The vertical wedges are used to determine the horizontal resolution of the receiver.

2. The circles are used to adjust the linearity of the picture.

3. The yoke, the focus control, and the centering adjustments.

4. The vertical-linearity and height controls.

5. The front-panel controls are those readily accessible to the viewer, such as the contrast, brightness, volume, etc. The preset controls are not readily accessible to the viewer and include the height, vertical-linearity, agc, etc.

Index

A

Action of horizontal differentiating circuit, 103-104
Active power filter, 34
Adjustment, deflection yoke, 182-183
Advantages of stacked arrays, 117
Afc antihunt network, 77
Aft defeat switch, 136
Agc
  amplified, 172-173
  delayed, 176-178
  forward, 130, 173
  keyed, 173-176
  rectified, 171-172
  reverse, 173
Allocations, channel, 88
Amplification, sync-pulse, 98-99
Amplified agc, 172-173
Amplifier
  reflex, 152
  rf, 128-135
  sync, 98
Amplitude modulation, 85
Anode, focus, 14, 21
Antenna
  cage, 116
  color laser, 122
  corner-reflector, 117-118
  in-line, 115
  log-periodic, 121
  rotators, 118
  systems, master, 122
  towers, 121
  V-type, 116
  Yagi, 114-115
Arrays, stacked, 117
Aspect ratio, 85
Assembly, electron gun, 15
Astigmatism, 16
Asymmetrical, or unbalanced multivibrator, 50
Audio detectors, 153-158
Automatic fine tuning, 135-136

B

Balun transformer, 123
Bandwidth and gain characteristics, 159

Basic elements of electromagnetic scanning system, 65
Basic transistor multivibrator, 54
Bass boost network, 160
Beam
  centering, 22-24
  control, 11-17
  deflection, 15-17, 22
  effect of fluorescent screen, 17
  focusing, 19-22
  formation, 11
Bifilar transformers, 142
Blacker than black, 89
Blanking level, 93
Blocking oscillator, 46, 52-54, 74-75
  vertical, 77-79
  waveforms in, 54
Boosted B+, 70
Bridged-T networks, 144-146
Brightness control, 166-167, 183
Broadband problem, 115-116

C

Cable, coaxial, 120
Cage antenna, 116
Capacitive tuning, 136
Capacitor
  charge and discharge paths in a multivibrator, 48
  neutralizing, 130
Cathode
  -coupled multivibrator, 50-52, 73-74
    pulse control of, 63
    waveshapes, 52
  follower, sync-separation circuits, 96
  ray tube, construction of, 11-15
Center impedance of half-wave dipole, 113
Centering
  beam, 22-24
  controls, 185
  rings, 24
Channel allocations, 88
Channels, uhf, 88
Characteristics, bandwidth and gain, 159
Chart, resolution, 179
Choke, filter, 35
Circuit(s)
  charging, RC, 39-40

Circuit(s)—cont
  discharge, RC, 40-41
  horizontal-oscillator, 72-77
  integrating, 101
  mixer, 131-135
  oscillator, 131
  RC differentiating, 100
Circular polarization, 106
Clamper, 98
Clipper, sync, 98
Clipping, sync-pulse, 98-99
Coaxial cable, 120
Co-channel interference, 108
Coil(s)
  deflection, 22-23
  horizontal waveform, 76
Color-laser antenna, 122
Combination series and shunt peaking, 163-164
Compensation, low-frequency, 159-161
Construction of cathode ray tube, 11-15
Contrast control, 167-168
Control(s)
  brightness, 166-167, 183
  centering, 185
  contrast, 167-168
  focus, 183
  front-panel, 180
  height, 185
  horizontal
    -drive, 69-70
    hold, 183
    -linearity, 70
  of beam, 11-17
  of scanning generators by sync pulses, 61-63
  preset, 180
  vertical
    hold, 183
    -linearity, 185-186
  width, 69, 185
Conventional multivibrator, 47-50
Converter, uhf, 136
Corner-reflector antenna, 117-118

D

Damper, 67-68
Dc restorer, 169
Deflection
  beam, 15-17, 22
  circuits
    horizontal, 67-71
    primary function of, 67
  coils, 22-23
  sensitivity, 16
  systems, vertical, 77-82
  -yoke
    adjustment, 182-183
    typical, 22
Degenerative traps, 144
Deionization potential, 45
Delayed agc for rf stage, 176-178
Delta sound detector, 157-158
Detection, slope, 157

Detector(s)
  audio, 153-158
  gated-beam, 155-156
  locked-oscillator, 156-157
  polarity, 147
  ratio, 153-155
  video, 147-149
Differentiation, 100
Differentiator voltage, 42-43
Diode
  sync-separation circuits, 94-96
  varactor, 128
  zener, 34
Dipole
  folded, 113-114
  half-wave, 111-113
  reception pattern, 112
Direct-current component of video signal, 90-92
Disc tuners, 126
Double-sideband modulation, 86
Dual gate MOSFET, 131

E

Effect of magnetic fields on electron beam, 19
Electromagnetic
  deflection, peaking circuits for, 65
  scanning system, basic elements, 65
  waves, 109
Electron
  beam, affected by magnetic fields, 19
  gun assembly, 15
Electronic filter, 34
Electrostatic focus, 14-15
Elements
  focus and accelerating controls, 12-15
  grid-control, 11-12
  parasitic, 114
Encapsulated parallel line, 119-120

F

Field, 85
Filter
  choke, 35
  electronic, 34
Fine tuning, automatic, 135-136
Firing point of sweep generator, 62
First anode, 14
Flat-ribbon parallel line, 119
Flyback high-voltage system, 71-72
Focus
  -and accelerating-control elements, 12-15
  anode, 14
  control, 183
Focusing
  anode, 21
  of beam, 19-22
Folded dipole, 113-114
Formation of
  beam, 11
  square and sawtooth waves, 42-43
Forward agc, 130, 173
Frame, 85

Frequencies, vhf, 105
Frequency
    adjustments, horizontal oscillator, 184-185
    synchronizing, 62
Free-running
    oscillator, pulse control of, 62
    frequency of generator, 61
Fringe area television reception, 121-122
Front-panel controls, 180
Function of vertical-equalizing pulses, 102-103

G

Gated
    -beam detector, 155-156
    sync-separation circuits, 97
Generator(s)
    sine-wave, 54
    test pattern, 179
Ghosts due to
    multipath transmission, 109-111
    reflections in the lead-in, 111
Grid-control element, 11-12
Ground wave, 107
Gun, electron, 14

H

Half-wave
    dipole, 111-113
        center impedance, 113
    rectifier, 32
Hartley oscillator, 76
Height control, 185
Hertzian doublet, 111
High-frequency
    compensation, 161-164
    phase shift, 165
High-voltage
    multiplication, 72
    power supply, 71-72
    system, flyback, 71-72
Horizontal
    -deflection circuits, 67-71
        solid-state, 70-71
    -drive control, 69-70
    hold, 183
    -linearity control, 70
    -oscillator
        circuits, 72-77
        frequency adjustment, 184-185
    polarization, 106
    -pulse separation, 100-101
    scanning waveform, 60
    waveform coil, 76

I

IC audio section, 153
Ideal
    phase shift, 165-166
    sawtooth, 59
If
    systems, sound, 151-153
    takeoff, sound, 151
Image orthicon, 25-26

In-line antenna, 115
Integrating
    circuit, 101
    network, vertical, 101
Integrator voltage, 42
Interference, co-channel, 108
Interlaced scanning, 102
Inverter, sync, 98
Ionization potential, 45
Ions, removal from electron beam, 24

K

Keyed agc, 173-176

L

Lead-in
    causes ghosts, 111
    types of, 118-120
Level, blanking, 93
Leveler, sync, 98
Limiter
    pulse, 98
    sync, 98
Lines
    pairing of, 61
    per frame, 85
Location of sync pulses, 93
Locked-oscillator detector, 156-157
Log-periodic antenna, 121
Low-frequency
    compensation, 159-161
    phase shift, 165

M

Master antenna systems, 122
Mixer circuits, 131-135
Modulating frequencies, television video signal, 85
Modulation
    amplitude, 85
    double-sideband, 86
    negative, 90
    single-sideband, 86
    vestigial-sideband video, 86-89
Monoscope, 28
MOSFET, dual gate, 131
Multipath transmission causes ghosts, 109-111
Multiplication, high-voltage, 72
Multivibrator(s), 46, 47-52
    and blocking oscillators, summary of, 56-57
    asymmetrical or unbalanced, 50
    cathode-coupled, 50-52, 73-74
    conventional, 47-50
    producing sawtooth scanning, 50
    vertical, 79-82

N

Negative modulation, 90
Neon-tube oscillator, 45
Network
    antihunt, 77
    bass-boost, 160
    bridged-T, 144-146

Neutralizing capacitor, 130
Neutrode rf amplifiers, 130
Noise problem, 109
Nuvistor rf amplifiers, 130

O

Obtaining wideband response in video if systems, 141-143
Orthicon, image, 25-26
Oscillator
  blocking, 46, 52-54, 74-75
  circuits, 131
  Hartley, 76
  neon-tube, 45
  relaxation, 45, 46
  sine-wave, 46
  thyratron, 46

P

Pairing of lines, 61
Parallel line
  encapsulated, 119-120
  flat-ribbon, 119
  tubular, 119
Parallel-tuned traps, 144
Parasitic element, 114
Pattern, test, 179
Peaking
  circuits
    for electromagnetic deflection, 65
    typical, 65
  series, 163
    and shunt combination, 163-164
  shunt, 162-163
Pentode sync-separation circuits, 96
Phase shift
  high-frequency, 165
  ideal, 165-166
  in video amplifier, 164-166
  low-frequency, 165
Picture
  elements per frame, 85
  smearing, 166
  tubes, scanning requirements, 63-65
Polarity of detector, 147
Polarization of transmitted wave, 105-106
Potential
  deionization, 45
  ionization, 45
Power
  filter, active, 34
  supplies, 32-36
  supply, high-voltage, 71-72
Preset controls, 180
Primary function of deflection circuits, 67
Pulse
  control of
    cathode-coupled multivibrator, 63
    free-running oscillator, 62
    vertical blocking oscillator, 62-63
  limiter, 98
  -width system, 75-76

R

Rabbit ears, 116
Ratio
  aspect, 85
  detectors, 153-155
RC
  charging curve, 40
  circuit
    charging, 39-40
    discharge, 40-41
    time constant, 41-42
  differentiating circuit, 100
  discharging curve, 40
Receiving antennas, factors to consider, 105
Reception pattern of dipole antenna, 112
Rectified agc, 171-172
Rectifier(s)
  half-wave, 32
  solid-state, 32-33
Reflectors and directors, 113
Reflex amplifier, 152
Regulation and filtering, 33-35
Regulator
  shunt, 34
  transistor, 33-34
Rejection of undesired adjacent-channel and co-channel
    carriers, 143-146
Relaxation oscillator, 45, 46
Removing ions from electron beam, 24
Resistor, shunting, 163
Response
  characteristics, video if, 140-141
  curve, video if, 140-141
Retrace lines, 168
Reverse agc, 173
RF amplifiers, 128-135
  neutrode, 130
  nuvistor, 130
  tetrode, 130
  transistor, 130-131
Rings, centering, 24
Rotators, antenna, 118

S

Saturation point, 81
Sawtooth
  generators, transistor, 54-56
    blocking oscillators, 55-56
    multivibrators, 54-55
    vacuum-tube, 46-54
  ideal, 59
  scanning, use of multivibrator, 50
Scanning, 28-29
  interlaced, 102
  requirements for picture tubes, 63-65
  waveform
    horizontal
    vertical, 60
Second anode, 14
Separation
  horizontal-pulse, 100-101

Separation—cont
   sync-pulse, 93-98
   vertical-pulse, 101-102
Series
   peaking, 163
   regulator, 33
   -tuned traps, 144
Shaping, sync-pulse, 98-99
Shielded twin-lead, 120
Shunt
   peaking, 162-163
   regulator, 34
Shunting resistor, 163
Sidebands, 85
Signal, video, 89-90
Sine-wave
   generators, 54
   oscillator, 46
Single-sideband modulation, 86
Sky wave, 107-108
Slope detection, 157
Smearing of picture, 166
Solid-state
   high voltage multiplier circuit, 72
   horizontal-deflection circuits, 70-79
   mixer circuit, 134-135
   rectifiers, 32-33
Sorting sync pulses, 99-102
Sound
   detector, delta, 157-158
   if
     systems, 151-153
     takeoff, 151
Square and sawtooth waves, formation of, 42-43
Stacked arrays, 117
   advantages of, 117
Standards for television broadcasts, 29
Summary of
   multivibrators and blocking oscillators, 56-57
   scanning-generator pulse control, 63
Sweep generator
   firing point, 62
   free-running frequency, 61
Switch-type tuner, 123-126
Sync
   amplifier, 98
   clipper, 98
   inverter, 98
   leveler, 98
   limiter, 98
   -pulse
     amplification, clipping, and shaping, 98-99
     clipping, 98-99
     control of scanning generator, 61-63
     location of, 93
     separation, 93-98
     shaping, 98-99
   -separation circuits
     cathode follower, 96
     diode, 94-96
     gated, 97
     pentode, 96

Sync—cont
   -separator
     circuits, transistor, 97-98
     triode
Synchronizing frequency, 62
System
   pulse-width, 75-76
   tuning, 123-128
   video if, 140-147

T

Takeoff, sound if, 151
Television
   broadcast standards, 29
   channels, wavelengths of, 108-109
   reception in fringe areas, 121-122
   resolution chart, 179
Test pattern, 179
   generator, 179
Tetrode rf amplifiers, 130
Thermal runaway prevention, 79
Thyratron oscillator, 46
Time constants of an RC circuit, 41-42
Trailer, 166
Transformer
   balun, 123
   bifilar, 142
Transistor
   oscillator circuit, 131
     regulator, 33-34
     rf amplifiers, 130-131
     sawtooth generators, 54-56
     blocking oscillators, 55-56
   sync-separator circuits, 97-98
Transmitted wave, polarization of, 105-106
Traps
   degenerative, 144
   parallel tuned, 144
   series-tuned, 144
Triode sync separators, 95
Tripler, voltage, 72
Tube operating conditions in a multivibrator, 48
Tubular parallel line, 119
Tuners
   disc, 126
   switch type, 123-126
   turret, 126
   uhf, 136-137
   varactor, 126-128
Tuning
   capacitive, 136
   methods, uhf, 136
   systems, 123-128
Turret tuners, 126
Twin lead, shielded, 120
Types of
   lead-in, 118-120
   wave paths, 106-108
Typical peaking circuit, 65

U

Uhf
   channels, 88

Uhf—cont
  converter, 136
  tuners, 136-137
  tuning methods, 136

V

Vacuum-tube sawtooth generators, 46-54
Varactor
  diode, 128
  tuners, 126-128
Vertical
  amplifier stage, 82
  blocking oscillator, 77-79
    pulse control, 62-63
  -deflection systems, 77-82
  -equalizing pulses, function of, 102-103
  hold, 183
  integrating network, 101
  -linearity control, 185-186
  multivibrator, 79-82
  -pulse separation, 101-102
  scanning waveform, 60
Vestigial-sideband video modulation, 86-89
Vhf frequencies, 105
Video
  amplifier, phase shift in, 164-166
  coupling to picture tube, 168-169
  detectors, 147-149
  if
    amplifiers
      with overcoupled transformers, 141-142
      with stagger-tuned circuits, 142-143
    response characteristics, 140-141
    systems, 140-147
      wideband response in, 141-143

Video—cont
  signal, 89-90
    direct-current component, 90-92
Vidicon, 26-27
Voltage
  differentiator, 42-43
  integrator, 42
  tripler, 72
  waveforms in a blocking oscillator, 54
V-type antenna, 116

W

Wave(s)
  electromagnetic, 109
  ground, 107
  paths, types of, 106-108
  sky, 107-108
Waveforms
  for transistor multivibrator circuit, 55
  of symmetrical multivibrator, 49
Wavelength, 109
  of television channels, 108-109
Waveshapes for cathode-coupled multivibrator, 52
White region, 89
Width control, 69, 185
  metal sleeve, 69
  variable inductor, 69

Y

Yagi array, 114-115

Z

Zener diode, 34